John Harry

**Introduction to
Plasma Technology**

Related Titles

Aliofkhazraei, M., Sabour Rouhaghdam, A.

Fabrication of Nanostructures by Plasma Electrolysis

approx. 280 pages with approx. 182 figures
2010
Hardcover
ISBN: 978-3-527-32675-4

Rauscher, H., Perucca, M., Buyle, G. (eds.)

Plasma Technology for Hyperfunctional Surfaces

Food, Biomedical, and Textile Applications

428 pages with 170 figures and 59 tables
2010
Hardcover
ISBN: 978-3-527-32654-9

Kawai, Y., Ikegami, H., Sato, N., Matsuda, A., Uchino, K., Kuzuya, M., Mizuno, A. (eds.)

Industrial Plasma Technology

Applications from Environmental to Energy Technologies

464 pages with 285 figures and 23 tables
2010
Hardcover
ISBN: 978-3-527-32544-3

Guest, G.

Electron Cyclotron Heating of Plasmas

264 pages with approx. 40 figures
2009
Hardcover
ISBN: 978-3-527-40916-7

Ochkin, V. N.

Spectroscopy of Low Temperature Plasma

630 pages with 227 figures and 80 tables
2009
Hardcover
ISBN: 978-3-527-40778-1

Ostrikov, K.

Plasma Nanoscience

Basic Concepts and Applications of Deterministic Nanofabrication

563 pages with 140 figures and 18 tables
2008
Hardcover
ISBN: 978-3-527-40740-8

Lieberman, M. A., Lichtenberg, A. J.

Principles of Plasma Discharges and Materials Processing

800 pages
2008
E-Book
ISBN: 978-0-470-36189-4

Heimann, R. B.

Plasma Spray Coating

Principles and Applications

449 pages with 236 figures and 19 tables
2008
Hardcover
ISBN: 978-3-527-32050-9

Hippler, R., Kersten, H., Schmidt, M., Schoenbach, K. H. (eds.)

Low Temperature Plasmas

Fundamentals, Technologies and Techniques

945 pages in 2 volumes with 498 figures
2008
Hardcover
ISBN: 978-3-527-40673-9

John Harry

Introduction to Plasma Technology

Science, Engineering and Applications

WILEY-VCH Verlag GmbH & Co. KGaA

The Author

Dr. John Harry
Highview
Knossington Road
Braunston
Oakham
Rutland LE15 8QX
United Kingdom

■ All books published by Wiley-VCH are carefully produced. Nevertheless, authors, editors, and publisher do not warrant the information contained in these books, including this book, to be free of errors. Readers are advised to keep in mind that statements, data, illustrations, procedural details or other items may inadvertently be inaccurate.

Library of Congress Card No.: applied for

British Library Cataloguing-in-Publication Data
A catalogue record for this book is available from the British Library.

Bibliographic information published by the Deutsche Nationalbibliothek
The Deutsche Nationalbibliothek lists this publication in the Deutsche Nationalbibliografie; detailed bibliographic data are available on the Internet at <http://dnb.d-nb.de>.

© 2010 WILEY-VCH Verlag & Co. KGaA, Boschstr. 12, 69469 Weinheim, Germany

All rights reserved (including those of translation into other languages). No part of this book may be reproduced in any form – by photoprinting, microfilm, or any other means – nor transmitted or translated into a machine language without written permission from the publishers. Registered names, trademarks, etc. used in this book, even when not specifically marked as such, are not to be considered unprotected by law.

Cover   Adam Design, Weinheim
Typesetting   Laserwords Private Limited, Chennai, India
Printing and Binding   Fabulous Printers Pte Ltd, Singapore

Printed in Singapore
Printed on acid-free paper

ISBN: 978-3-527-32763-8

Contents

Preface *XI*
Symbols, Constants and Electronic Symbols *XIII*

1	**Plasma, an Overview** *1*	
1.1	Introduction *1*	
1.2	Plasma *4*	
1.2.1	Space Plasmas *4*	
1.2.2	Kinetic Plasmas *4*	
1.2.3	Technological Plasmas *5*	
1.3	Classical Models *5*	
1.3.1	Simple Ballistic and Statistical Models *5*	
1.3.2	Statistical Behaviour *6*	
1.3.3	Collisions Between Particles *8*	
1.3.4	Coulomb Forces *9*	
1.3.5	Boundaries and Sheaths *10*	
1.3.6	Degree of Ionization *10*	
1.4	Plasma Resonance *11*	
1.5	The Defining Characteristics of a Plasma *11*	
	References *13*	
	Further Reading *13*	
2	**Elastic and Inelastic Collision Processes in Weakly Ionized Gases** *15*	
2.1	Introduction *15*	
2.2	The Drift Velocity *15*	
2.2.1	Electrical Conductivity *17*	
2.2.2	Mobility *17*	
2.2.3	Thermal Velocity *18*	
2.2.4	Collision Frequency *18*	
2.2.5	Collision Cross-section *19*	
2.3	Inelastic Collision Processes *21*	
2.3.1	Excitation *22*	
2.3.1.1	Metastable Processes *22*	

Introduction to Plasma Technology: Science, Engineering and Applications. John Harry
Copyright © 2010 WILEY-VCH Verlag GmbH & Co. KGaA, Weinheim
ISBN: 978-3-527-32763-8

2.3.2	Ionization and Recombination Processes	*23*
2.3.2.1	Charge Transfer	*24*
2.3.2.2	Dissociation	*24*
2.3.2.3	Negative Ionization	*24*
2.3.2.4	Recombination	*24*
2.3.2.5	Metastable Ionization	*25*
	References	*26*
3	**The Interaction of Electromagnetic Fields with Plasmas**	*29*
3.1	Introduction	*29*
3.2	The Behaviour of Plasmas at DC and Low Frequencies in the Near Field	*29*
3.2.1	Charged Particles in Electromagnetic Fields	*31*
3.2.1.1	Behaviour of a Charged Particle in an Oscillating Electric Field	*32*
3.2.1.2	Plasma Frequency	*34*
3.2.1.3	The Debye Radius	*35*
3.3	Behaviour of Charged Particles in Magnetic Fields (Magnetized Plasmas)	*37*
3.4	Initiation of an Electrical Discharge or Plasma	*41*
3.5	Similarity Conditions	*41*
	References	*43*
	Further Reading	*43*
4	**Coupling Processes**	*45*
4.1	Introduction	*45*
4.2	Direct Coupling	*45*
4.2.1	The Cathode	*49*
4.2.1.1	Emission Processes	*51*
4.2.2	The Cathode Fall Region	*56*
4.2.3	The Anode	*57*
4.2.4	The Discharge Column	*57*
4.2.5	Interaction of Magnetic Fields with a Discharge or Plasma	*59*
4.3	Indirect Coupling	*62*
4.3.1	Induction Coupling	*62*
4.3.2	Capacitive Coupling	*64*
4.3.3	Propagation of an Electromagnetic Wave	*65*
4.3.4	The Helical Resonator	*68*
4.3.5	Microwave Waveguides	*69*
4.3.6	Electron Cyclotron Resonance	*70*
4.3.7	The Helicon Plasma Source	*74*
	References	*75*
	Further Reading	*75*

5	**Applications of Nonequilibrium Cold Low-pressure Discharges and Plasmas** *77*	
5.1	Introduction *77*	
5.2	Plasma Processes Used in Electronics Fabrication *77*	
5.2.1	The Glow Discharge Diode *80*	
5.2.2	The Magnetron *83*	
5.2.3	Inductively Coupled Plasmas *84*	
5.2.4	Electron Cyclotron Resonance Reactor *85*	
5.2.5	The Helical Reactor *86*	
5.2.6	The Helicon Reactor *87*	
5.3	Low-pressure Electric Discharge and Plasma Lamps *88*	
5.3.1	The Low-pressure Mercury Vapour Lamp *88*	
5.3.2	Cold Cathode Low-pressure Lamps *91*	
5.3.3	Electrodeless Low-pressure Discharge Lamps *91*	
5.4	Gas Lasers *91*	
5.5	Free Electron and Ion Beams *94*	
5.5.1	Electron and Ion Beam Evaporation *94*	
5.5.2	Ion Beam Processes *95*	
5.5.3	High-power Electron Beams *97*	
5.6	Glow Discharge Surface Treatment *99*	
5.7	Propulsion in Space *100*	
	References *101*	
	Further Reading *101*	
6	**Nonequilibrium Atmospheric Pressure Discharges and Plasmas** *103*	
6.1	Introduction *103*	
6.2	Atmospheric Pressure Discharges *103*	
6.2.1	Corona Discharges *105*	
6.2.2	Corona Discharges on Conductors *108*	
6.3	Electrostatic Charging Processes *110*	
6.3.1	Electrostatic Precipitators *110*	
6.3.2	Electrostatic Deposition *113*	
6.4	Dielectric Barrier Discharges *114*	
6.5	Plasma Display Panels *116*	
6.6	Manufacture of Ozone *116*	
6.7	Surface Treatment Using Barrier Discharges *118*	
6.8	Mercury-free Lamps *118*	
6.9	Partial Discharges *118*	
6.10	Surface Discharges *120*	
	Further Reading *121*	
7	**Plasmas in Charge and Thermal Equilibrium; Arc Processes** *123*	
7.1	Introduction *123*	
7.2	Arc Welding *124*	
7.2.1	Metal Inert Gas Welding *126*	

7.2.2	Tungsten Inert Gas Welding	*127*
7.2.3	Submerged Arc Welding	*129*
7.2.4	The Plasma Torch	*129*
7.3	Electric Arc Melting	*131*
7.3.1	The Three-phase AC Arc Furnace	*131*
7.3.2	DC Arc Furnaces	*134*
7.3.3	Electric Arc Smelting	*135*
7.3.4	Plasma Melting Furnaces	*136*
7.3.5	Vacuum Arc Furnaces	*137*
7.4	Arc Gas Heaters	*138*
7.4.1	Inductively Coupled Arc Discharges	*139*
7.5	High-pressure Discharge Lamps	*141*
7.6	Ion Lasers	*144*
7.7	Arc Interrupters	*145*
7.7.1	Vacuum Circuit Breakers and Contactors	*147*
7.8	Magnetoplasmadynamic Power Generation	*149*
7.9	Generation of Electricity by Nuclear Fusion	*149*
7.10	Natural Phenomena	*150*
7.10.1	Lightning	*150*
	Further Reading	*152*
8	**Diagnostic Methods**	*155*
8.1	Introduction	*155*
8.2	Neutral Particle Density Measurement	*155*
8.3	Probes and Sensors	*156*
8.3.1	The Langmuir Probe	*156*
8.3.2	Magnetic Probes	*158*
8.4	Optical Spectroscopy	*159*
8.4.1	Optical Emission Spectroscopy	*159*
8.4.2	Absorption Spectroscopy	*161*
8.4.3	Scattering Measurements	*161*
8.5	Interferometry	*162*
8.5.1	Microwave Interferometer	*163*
8.6	Mass Spectrometry	*164*
8.7	Electrical Measurements	*165*
8.7.1	Electrical Instrumentation	*166*
8.7.2	The Oscilloscope	*167*
8.7.3	Electrical Measurements Using Probes	*168*
8.7.4	Current Measurement	*170*
	Further Reading	*172*
9	**Matching, Resonance and Stability**	*173*
9.1	Introduction	*173*
9.2	The Plasma Characteristic	*173*
9.3	Stabilizing Methods	*176*

9.3.1	Reactive Stabilization *176*	
9.4	Effect of Frequency *179*	
9.5	Interaction between the Plasma and Power Supply Time Constants *179*	
9.6	Matching *180*	
9.7	Resonance *182*	
9.8	Parasitic Inductance and Capacitance *183*	
	Further Reading *185*	
10	**Plasma Power Supplies** *187*	
10.1	Introduction *187*	
10.2	Transformers and Inductors *187*	
10.3	Rectification *191*	
10.4	Semiconductor Power Supplies *193*	
10.4.1	The Inverter Circuit *193*	
10.4.2	Semiconductor Switches *195*	
10.4.3	Current Control *195*	
10.4.4	The Inverter Circuit *196*	
10.4.5	Converter Circuits *197*	
10.4.6	Inverter Frequencies *198*	
10.4.7	High-Frequency Inverter *198*	
10.5	Electronic Valve Oscillators *199*	
10.6	Microwave Power Supplies *199*	
10.7	Pulsed Power Supplies *200*	
10.8	Ignition Power Supplies *201*	
10.9	Electromagnetic Interference *205*	
10.9.1	Conduction *206*	
	Further Reading *207*	

Index *209*

Preface

Plasma plays an ever increasing role in industrial, commercial and domestic environments and also space and fusion research. The evolution of new applications of plasmas continues to accelerate at an increasing rate. Applications in medicine, textile treatment, solar cells, electrohydraulic water treatment, paper, packaging and corrosion protection are among many.

Early books have treated the topic as a whole and it was within the curriculum of many undergraduate courses. However, since the replacement of the electronic valve by semiconductors and with the increasing complexity of the subject, books have become more specialized and application orientated. This has impeded an introduction to plasma technology and limited knowledge of plasma processes and applications and opportunities for cross-fertilization have been missed. Those in the field are often unaware of the different methods for producing plasmas and applying plasma technology outside their own area of expertise. Indeed, the apparent complexity of the subject has restricted practitioners largely to physicists and chemists at graduate or postgraduate level and the subject is regarded as opaque.

A fundamental challenge is the need to couple energy into a plasma. When plasmas were limited to DC and low-frequency AC this was relatively straightforward. Coupling energy into a plasma, particularly at high frequencies and low gas pressures, has become a critical area. Few books today give power supplies any more than cursory attention, when in fact energization of the plasma is at the core of the process and the selection, matching and correct operation of the power supply are critical to the success of research or industrial processes.

This book not only serves to fill the gaps that existing publications leave, but also develops an understanding of both the plasma and its interaction with the supply, which is essential both in research and to optimize applications. Recent advances in the use of semiconductors to generate power at high frequencies and the development of high-speed switching methods have enabled complex high-frequency electronic supplies to be developed, opening up many new areas of application such as in medicine and textile treatment. This book addresses the problem of design and selection, matching and optimizing the power supply for a given process.

Introduction to Plasma Technology: Science, Engineering and Applications. John Harry
Copyright © 2010 WILEY-VCH Verlag GmbH & Co. KGaA, Weinheim
ISBN: 978-3-527-32763-8

The objective of this book is to make the subject accessible, and this is achieved by providing a concise, unified introduction to the subject over the full range of plasma operation at a level appropriate to professional engineers or scientists, final-year undergraduates in a technical discipline and postgraduates entering the field and as a reference text.

The philosophy of the book is to treat the subject in the simplest of terms so that a clear understanding is achieved using simple models without resorting to complex mathematics. Varying degrees of complexity are developed by superimposing the different processes in the same way as superposition is used to solve problems in electric circuits.

Finally, I would like to thank David Hoare for his patient help, advice and insight in helping me with writing the book, Ben Thompson for his painstaking creation of the diagrams and my wife Suzanne for her encouragement and patience and willing assistance with proofreading.

Oakham, Rutland, UK
March 2010

John Harry

Symbols, Constants and Electronic Symbols

Symbols

a	acceleration
a	radius
A	area
B	magnetic field
c	velocity of light
d	distance
d_s	thickness of plasma sheath
D	diffusion coefficient
D	Debye length
e	electron charge
E	electric field
$f(w)$	energy distribution function
$f(u)$	velocity distribution function
f_{ce}	critical electron frequency
f_{ci}	critical ion frequency
h	Planck's constant
i	instantaneous current
I	current
J	current density
k_B	Boltzmann's constant
L	characteristic reactor dimension
m, M	mass of particle
n	particle number density
n_D	number of particles in a Debye sphere
p	pressure
P	power
q	electrical charge
r	radius
s	distance
t	time
T	temperature

Introduction to Plasma Technology: Science, Engineering and Applications. John Harry
Copyright © 2010 WILEY-VCH Verlag GmbH & Co. KGaA, Weinheim
ISBN: 978-3-527-32763-8

u, v	velocity
v	instantaneous voltage
V	electrical potential, voltage
Vol	volume
W	energy
α	degree of ionization
ε_0	permittivity of free space
ε_r	relative pemittivity
μ	mobility
μ	refractive index
ω	angular frequency
τ	residence time
δ	skin depth
E	energy
ϕ	phase angle
ϕ_w	work function (eV)
λ	wavelength
λ_e	mean free path electron
λ_i	mean free path ion
λ_D	Debye length
μ_0	permeability of free space
μ_r	relative permeability
ν	collision frequency
ν_e	electron collision frequency
ν_i	ion collision frequency
θ	angle
ρ	electrical resistivity
σ	electrical conductivity
η	viscosity
τ	period of frequency
ω	angular frequency
ω_g	gyro frequency
ω_e	electron angular frequency
ω_i	ion angular frequency
ω_c	cyclotron frequency
ω_{pe}	electron plasma angular resonant frequency
ω_{pi}	ion plasma angular resonant frequency
ω_s	supply angular frequency

Useful Constants

Charge of electron, e, 1.6×10^{-19} C
Avogadro's number, n_A, 6.02×10^{23} particles mol^{-1} at NTP (normal temperature and pressure, 20 °C and 760 Torr)
Boltzmann's constant, $k = R_0/n_A$, 1.38×10^{-23} J K^{-1}
Electronvolt, eV, 1.6×10^{-19} J
Loschmidt's number, n_L, 2.69×10^{25} particles m^{-3} in a gas at NTP
Mass of electron, m_e, 9.11×10^{-31} kg
Mass of proton (hydrogen atom), m_i, 1.67×10^{-27} kg
Mean free path (nitrogen), λ_h, 6.63×10^{-8} m at NTP
Permeability of free space μ_0, $4\pi \times 10^{-7}$ H m^{-1}
Permittivity of free space ε_0, 8.85×10^{-12} F m^{-1}
Planck's constant, h, 6.626×10^{-34} J s
Random velocity (nitrogen molecule), u_r, 509 m s^{-1} at NTP
Ratio of mass of proton to mass of electron, 1833
Universal gas constant, R_0, 8.31 J K^{-1} mol^{-1}
Stefan–Boltzmann constant, k_b, 5.67×10^{-8} W m^{-2} K^{-4}
Velocity of light, c_0, 3×10^8 m s^{-1}
1 bar = 760 Torr 101 kPa

Pressure Units Conversion

1 bar = 760 Torr = 100 kPa or 10^5 Pa
1 mbar = 100 Pa
1 Torr = 133 Pa
5 Torr = 665 Pa
1 m Torr = 0.133 Pa
101 kPa = 760 Torr = 1 atm = 10^5 Pa
1 kPa = 7.52 Torr
100 Pa = 0.752 Torr
1 Pa = 7.52×10^{-3} Torr
1 mPa = 7.52×10^{-6} Torr

Some useful plasma relationships

Electron plasma frequency $\omega_{pe} = \left(\frac{n_e e^2}{m_e \varepsilon_0}\right)^{\frac{1}{2}}$

Ion plasma frequency $\omega_{pi} = \left(\frac{n_i e^2}{m_i \varepsilon_0}\right)^{\frac{1}{2}}$

Electron cyclotron or gyro frequency $\omega_{ce} = \frac{eB}{m_e}$

Ion cyclotron or gyro frequency $\omega_{ci} = \frac{eB}{m_i}$

Debye length $\lambda_D = \left(\frac{k_B T_e \varepsilon_0}{n_e e^2}\right)^{\frac{1}{2}}$

Free electrons in Debye volume $N_D = \frac{4}{3}\pi n_e \lambda_D^3$

Velocity of electro-magnetic waves $v = \frac{1}{\sqrt{\mu_0 \mu_r \varepsilon_0 \varepsilon_r}}$

Velocity of light in free space $c = \frac{1}{\sqrt{\mu_0 \varepsilon_0}}$

Impedance of free space $Z_0 = \sqrt{\frac{\mu_0}{\varepsilon_0}}$

Phase velocity $v_p = \frac{\omega}{k}$

Group velocity $v_g = \frac{d\omega}{dk}$

Electronic Symbols

Symbol	Name
	air cored transformer
	amplifier
	capacitor
	diode
	discharge
	earth connection
	ferrite ring inductor
	IGBT
	inductor
	iron cored transformer
	MOSFET
	resistor
	thyristor

1
Plasma, an Overview

1.1
Introduction

This chapter introduces the different areas of plasma, the unique aspects of the subject, definitions, the use of simple ballistic and statistical models and the defining characteristics of plasmas.

The influence of plasma technology has penetrated almost every aspect of human activity during the last few years. Some of the different areas of plasma technology, applications and areas of operation are shown in Table 1.1. Despite the widespread use of many of the applications, the subject of plasma has developed a mystique which has given it a reputation of being complex and impenetrable. Aspects of plasmas which make the subject different from many other areas of physics and engineering are introduced in this chapter.

Plasma comprises, in its simplest form, the two elementary particles that make up an atom: electrons and ions. Over 99% of the universe is believed to be plasma, as opposed to condensed matter (solids, liquids and gases) such as comets, planets or cold stars. The term *plasma* was first used by Langmuir in 1927 and derives its name from the Greek to shape or to mould and the analogy with biological plasma, which is an electrolyte, and describes the self-regulating behaviour of plasma in contrast to the apparently random behaviour of fluids.

The science of plasma encompasses space plasmas, kinetic plasmas and technological plasmas and ranges over enormous variations of parameters such as pressure, distance and energy. One method of distinguishing different areas of plasma technology that is often used is as hot or cold plasmas (Table 1.2) depending on the relative value of the ion temperature T_i to the electron temperature T_e. Although widely and conveniently used to describe individual areas, they accentuate the differences, and the anomaly of a plasma at several thousand degrees kelvin being described as cold is not always helpful! Other common descriptions used are glow, corona, arc and beams. These artificial definitions often present obstacles to those entering the field or to those already engaged in it. The subject of plasma is better described as a continuum in terms primarily of the potential energy of electrons T_e and ions T_i and the electron number density n_e, and one of the objectives of this book is to emphasize the similarities rather than the differences.

Introduction to Plasma Technology: Science, Engineering and Applications. John Harry
Copyright © 2010 WILEY-VCH Verlag GmbH & Co. KGaA, Weinheim
ISBN: 978-3-527-32763-8

Table 1.1 Some applications of plasmas.

Low-pressure non-equilibrium plasma	Atmospheric non-equilibrium plasmas	High-current equilibrium plasmas
Plasma processes used in electronics fabrication	Atmospheric pressure discharges	Arc welding
Glow discharge diode	Corona discharges	Metal inert gas welding
Magnetron	Corona discharges on power lines	Tungsten inert gas welding
Induction coupled plasmas	Electrostatic charging processes	Submerged arc welding
Electron cyclotron resonance reactor	Electrostatic precipitators	Plasma torch
Helical reactor	Electrostatic deposition	Electric arc melting
Helicon reactor	Dielectric barrier discharges	Three-phase AC arc furnace
Low-pressure electric discharge and plasma Lamps	Manufacture of ozone	DC arc furnaces
Low-pressure mercury vapour lamp	Surface treatment using barrier discharges	Electric arc smelting
Cold cathode low-pressure lamps	Partial discharges	Plasma melting furnaces
Electrodeless low-pressure discharge lamps	Surface discharges	Vacuum arc furnaces
Plasma display panels	Atmospheric pressure glow discharges	Arc gas heaters
Gas lasers	Surface treatment of films and textiles to change surface properties	Electric discharge augmented fuel flames
Free electron and ion beams	Sterilization of medical equipment and disinfection	Induction coupled arc discharges
Electron and ion beam evaporation	Surgery	High-pressure discharge lamps
Ion beam processes	Diesel exhaust treatment	Ion lasers
High-power electron beams	Biomedical applications	Arc interrupters
Glow discharge surface treatment	Surface modification to improve adhesion, hydrophobic properties, wetting	Vacuum circuit breakers and contactors
Propulsion in space		Magnetoplasmadynamic power generation
		Generation of electricity by nuclear fusion
		Natural phenomena
		Lightning
		Applications in space

The reason for plasmas' unique characteristics and relevance to high-energy processes is apparent from Figure 1.1, where the electron temperature T_e is shown for different plasma processes as a function of electron number density of the electrons. Energy and temperature are related by the Boltzmann constant, k_B:

$$\frac{1}{2}mu^2 = k_B T$$

where $k_B = 1.38 \times 10^{-23}$ J K^{-1} [2]. In a cold plasma such as a neon lamp, the kinetic energy equates almost entirely to the electron energy and, although the mean electron temperature may be several times room temperature, the number of hot electrons is only a tiny fraction of the total and their thermal mass is small

Table 1.2 Temperature and pressure ranges of hot and cold plasmas.

Low-temperature thermal cold plasmas	Low-temperature non-thermal cold plasmas	High-temperature hot plasmas
$T_e \approx T_i \approx T < 2 \times 10^4$ K	$T_i \approx T \approx 300$ K $T_i \ll T_e \leqslant 10^5$ K	$T_i \approx T_e > 10^6$ K
Arcs at 100 kPa	Low pressure ~100 Pa glow and arc	Kinetic plasmas, fusion plasmas

From Ref. [1].

compared with an atom or molecule, so that the average temperature increase is small. The potential and energy of most plasma processes are several orders of magnitude greater than those of most other processes; for example, the energy of molecules at room temperature is about 0.025 eV and at 4000 K it is 0.35 eV.

It is usually necessary to consider plasma parameters on an atomic level, including particle sizes, length and time scales, particle number densities, forces between particles and many other parameters, with respect to each other. Atomic particles resonate and the difference in resonant frequencies between electrons and ions due to their different masses adds a further layer of complication, together with the fluid nature of a plasma and the ability to affect it with static and fluctuating

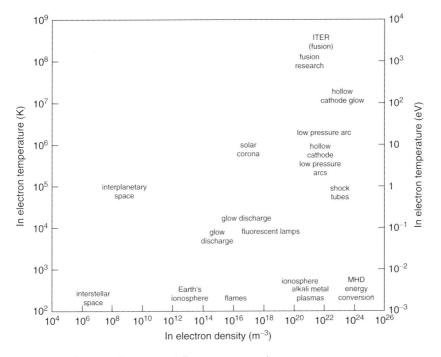

Figure 1.1 Plasma applications at different currents and gas pressures.

1.2
Plasma

1.2.1
Space Plasmas

Space plasmas [3] vary from very hot ($T > 30\,000$ K), dense plasmas at the centre of stars, corona flares and sunspots, to cold, less dense plasmas such as the aurora borealis and the ionosphere within the Earth's gravitational system.

Space is not a perfect vacuum but the gas pressure in interstellar space may be as low as about 3 fPa (22.5×10^{-18} Torr), at which pressure has little meaning. [The parameters of neutral particles scale approximately with number density (gas pressure) over wide ranges.] The corresponding particle density in interstellar space may be as low as 10^6 m^{-3}, less than 1×10^{-19} of the particle density at atmospheric pressure on Earth. The diameter of a hydrogen atom is of the order of 10^{-10} m with a nucleus of diameter about 10^{-15} m; the chance of a collision is very low, and electrons and ions travel through space at high velocities over large distances.

The fundamental theory of space plasmas is the same, however, as in other areas of plasma technology, although the conditions such as pressure, boundaries and energies may be very different. Space propulsion shares terrestrial technology such as plasma torches, electron cyclotrons or helicons, also used for making computer chips.

1.2.2
Kinetic Plasmas

Kinetic plasmas are generally described as hot plasmas since the ion temperature which is approximately equal to the electron temperature ($T_e \approx T_i$), is high although the gas is not necessarily in thermal equilibrium since the neutral atoms and molecules may be at a much lower temperature. In a kinetic plasma, the mean free path of a particle (λ) is long (i.e. the time between collisions τ is long and the collision frequency ν is low), electrons and ions tend to behave separately and their behaviour can be described in terms of individual particles in both space and time. Beams of electrons and ion beams at low pressures are used in semiconductor manufacture and for welding and melting can be regarded as kinetic plasmas.

At higher pressures, such as those used in atomic fusion [4], although the collision processes can be described as kinetic, the effects of diffusion gradients, collisions and the fluid and electromagnetic properties also affect the process, and they may also be described as magnetoplasmadynamic (MPD) [5]. Very high energy densities are possible and kinetic plasmas are the subject of areas such as fusion and particle research.

1.2.3
Technological Plasmas

Technological plasmas are normally supplied with energy from electric power sources, although excitation from shock waves and chemical plasmas is possible, and operate in the region from atmospheric pressure down to about 10 Pa (75.2×10^{-3} Torr) [6]. Gas pressures as low as 10^{-11} Pa (7.52×10^{-14} Torr) are obtainable in the laboratory, but the use of plasmas is limited by the energy density at low pressures to about 100×10^{-3} Pa (0.752×10^{-3} Torr), at which the mean free path is of the order of 100 mm.

Technological plasmas are often referred to as *cold plasmas*, since the neutral particle and ion temperatures are often much lower than the electron temperature. If the length and time scales of changes of the electric field are long compared with the mean free path and collision frequency, the effects of collisions in a plasma result in a statistical distribution of velocities and energy. At gas pressures above about 0.133 Pa (10^{-3} Torr), corresponding to particle densities of 10^{19} m^{-3}, plasmas can be considered statistically as a quasi-continuous fluid; below this there is a substantial separation between particles and the behaviour is more accurately described by the behaviour of individual particles (free electrons or ions), such as those in electron and ion beams.

1.3
Classical Models

1.3.1
Simple Ballistic and Statistical Models

It is difficult to comprehend the complexity of the numerous subatomic particles and their interactions, but fortunately simple models adequately explain the behaviour of most plasma processes.

Models using ballistic equations to describe atoms and electrons are a simple way of understanding the events that occur in a plasma at an atomic level. Many plasma processes can be treated using classical mechanics, such as the velocity equations:

$$s = ut + \frac{1}{2}ft^2, \quad v^2 = u^2 + 2fs$$

momentum equations:

$$m_1 u_1 = m_2 v_2$$

conservation of energy:

$$\frac{1}{2} m_1 u^2 = \frac{1}{2} m_2 v^2$$

the continuity equation and basic electromagnetic theory [7].

The atoms or molecules comprising a gas are in a continual state of movement due to thermal diffusion caused by temperature differences in the gas. Thermal diffusion affects the behaviour of plasmas, except at very low pressures, and it is necessary to superimpose a statistical model on the ballistic model using mean values and probabilities. The transport properties of the parent gas (density, velocity, viscosity and pressure) have a major effect on the plasma where the degree of ionization is low and the number of collisions is high.

The effects of the large numbers of collisions drive the number distribution of velocities towards a statistical distribution such as a Maxwellian distribution (see Section 1.3.2). A kinetic solution that describes the velocity and position as a function of time is appropriate where the time scales of significant functions such as waves, propagation, instabilities and other non-Maxwellian effects are much shorter than the time for relaxation or thermal equilibrium. Particle beams, some low-pressure plasmas and, for example, toroidal (tokomak) fusion reactors fall into this area.

1.3.2
Statistical Behaviour

Where there are a very large number of particles in thermal and charge equilibrium in an isotropic (uniform) medium, the velocity distribution can be determined statistically and is given by the Maxwell equation [2]:

$$f(u) = \frac{dn_u}{du} = \frac{4n}{\pi^{\frac{1}{2}}} \left(\frac{m}{2k_B T}\right)^{\frac{3}{2}} u^2 \exp\left(-\frac{mu^2}{2k_B T}\right) \tag{1.1}$$

where T is the average temperature and u the velocity over the range du, n is the particle number density, m the particle mass, and k_B Boltzmann's constant, $k_B = 1.38 \times 10^{-23}$. The kinetic energy of a particle is

$$\frac{mu^2}{2} = k_B T \tag{1.2}$$

The Maxwell velocity distribution (Figure 1.2) relates the normalized number probability of neutral particles which have a specific value of velocity [7].

The most probable velocity is the maximum of the velocity distribution:

$$u_m = \left(\frac{2k_B T}{m}\right)^{\frac{1}{2}} \tag{1.3}$$

The average velocity is

$$\bar{u}_{av} = \left(\frac{8k_B T}{\pi m}\right)^{\frac{1}{2}} \tag{1.4}$$

and the r.m.s. velocity is

$$\bar{u}_{rms} = \left(\frac{3k_B T}{m}\right)^{\frac{1}{2}} \tag{1.5}$$

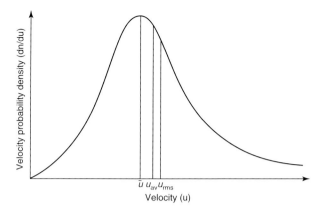

Figure 1.2 Maxwellian velocity distribution showing average, r.m.s. and peak values of probable velocities.

Although there is a wide distribution of velocities, nearly 90% are between half and double the average velocity and the probability of a molecule having a very high or very low velocity is extremely small; however, small numbers of high-velocity particles with high kinetic energy exist in the tail of the distribution.

The Maxwell–Boltzmann energy distribution $f(w)$ [1] (Figure 1.3) as a function of the probable particle distribution derived from the velocity distribution is

$$f(\mathbf{w}) = \frac{dn_w}{dw} = \frac{2n}{\pi^{\frac{1}{2}}} \frac{w^{\frac{1}{2}}}{(2k_BT)^{\frac{3}{2}}} \exp\left(-\frac{w}{2k_BT}\right) \tag{1.6}$$

The velocity distribution of charged particles tends to be non-Maxwellian, due to the effect of the electric field and neighbouring charged particles. However, it may be applied with caution in weak electric fields:

1) The electric field strength is low enough that inelastic collisions can be ignored but high enough for $T_e \gg T_i$.

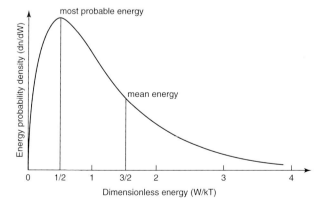

Figure 1.3 Maxwell–Boltzmann energy distribution.

2) The supply frequency is much lower than the collision frequency.
3) The collision frequency is independent of the electron energy T_i.

Figure 1.4 shows the energy levels for different atomic and molecular transitions. The energy gained by an electron (charge 1.6×10^{-19} C) accelerated through a potential of 1 V is 1 eV or 1.6×10^{-19} J. The Maxwell–Boltzmann distribution indicates the small number of particles available with sufficient energy for ionization.

One electronvolt corresponds to an average velocity of a proton of 1.38×10^4 m s^{-1} and for an electron 593×10^6 km s^{-1}, or a temperature of 11 600 K. The units of energy in electronvolts are small but the overall energy transfer may be large since the numbers and collision frequency are very high. On a molecular level, the energy required to change a molecule of a solid to a liquid is about 10^{-2} eV, but to ionize a gas molecule requires at least 1 eV. At room temperature, the kinetic energy of a proton (the nucleus of a hydrogen atom) is about 0.04 eV and only 0.5 eV at 4000 K. The energy required to detach an electron from a single proton to ionize it is 13.6 eV, so that the probability of a neutral particle having a high enough energy to ionize even at high temperatures is very low.

1.3.3
Collisions Between Particles

Collisions between particles may be elastic, in which case momentum is preserved, or inelastic, where momentum is transferred to potential energy; in both cases, energy is exchanged. The energy per collision and the collision frequency between atomic particles affect the rate at which energy can be coupled with or released from a plasma. The mass of an electron, m_e, is 9.11×10^{-31} kg. The atomic weight of the nucleus of a hydrogen atom (proton), m_p, is 1.67×10^{-27} kg.

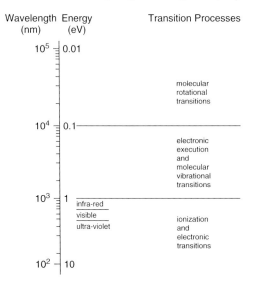

Figure 1.4 Atomic and molecular energy associated with different energy transitions.

The small mass of the electron results in only a negligible exchange of kinetic energy from electrons in an elastic collision and both particles preserve the magnitude of their momentum but its direction will be changed.

The fraction of energy k transferred in an elastic collision from a particle of mass m to a stationary particle of mass M can be derived from the momentum and kinetic energy equations. For impact on the axis of the two particles

$$k = \frac{\frac{4m}{M}}{\left(1 + \frac{m}{M}\right)^2} \tag{1.7}$$

Where $m \gg M$ such as an electron neutral collision

$$k \approx \frac{4m}{M} \tag{1.8}$$

If m and M are similar, the energy is approximately equally shared. In a collision between an electron and a stationary hydrogen atom or proton, $k = 2\,m/M$ and the energy transferred is 10.9×10^{-4} or about 0.1% of the initial electron energy; if the collision is inelastic, the fraction of energy converted may be as high as $M/(m+M)$, and over 99% of the initial electron energy may be transferred.

The collision frequency of a proton in air at atmospheric pressure is of the order of 7×10^9 s^{-1}. The mean free path λ (distance between collisions) of an atom or molecule in air at atmospheric pressure is approximately 6×10^{-8} m and decreases approximately inversely with pressure, so that at 0.1 Pa (0.752×10^{-3} Torr), the mean free path is of the order of 60 mm [8].

1.3.4
Coulomb Forces

The Coulomb force is the electrostatic force between electric charges; opposite charges attract, like charges repel. Electrons can be considered as point charges; ions may also be considered in the same way by assuming that their charge is concentrated at their centre.

The electrostatic force between two point charges q_1 and q_2 separated by a distance r is given by Coulomb's law [2] as

$$\frac{q_1 q_2}{4\pi r^2 \varepsilon_0} \tag{1.9}$$

For two charges of opposite polarity, the particles attract although forces due to other charged particles may prevent collision. The Coulomb force between particles in close proximity is strong and may be much greater than the effect of the applied electric field.

Displacement of a charge from a position of charge equilibrium results in oscillation of the charge at the electron or ion resonant frequency (see Section 1.4). When a large number of collisions occur at high pressure and an electron charge of depth equal to the Debye radius λ_D develops, this screens the effect of the electric field and reduces its potential (see Section 3.2.1.3). Ambipolar diffusion, in which the drift velocities of electrons and ions are affected by the electric field between them, retards high-velocity electrons. At low pressures and, for example, in electron and ion beams an external electric field may have a greater effect than Coulomb forces.

1.3.5
Boundaries and Sheaths

Technological plasmas have boundaries defined by the walls of a vessel, electrodes or the ambient gas and energized particles give up energy by collisions at the boundaries. Insulated surfaces or boundaries rapidly acquire a charge, forming a sheath, and repel like charges. Figure 1.5 illustrates the variation of plasma potential in a plasma sheath with distance from a boundary.

A local charge concentration of positive ions forms over a region of a few tens of microns comparable to the mean free path of the particles. The ion sheath causes a potential gradient between the plasma and the boundary and screens the plasma, which remains approximately charge neutral. The ability of the sheath to screen the plasma from a disturbance is known as the *Debye radius* [1]. The Debye radius is normally small compared with the principal dimensions of a process except at very high values of T_e or very low values of n_e except for example on a nanometre scale (as in etching and similar processes).

1.3.6
Degree of Ionization

The degree of ionization [1] is a measure of the number of ionized atoms or molecules (which is normally equal to the number of electrons n_e or ions n_i) as a fraction α of the total number n_t of atoms or molecules:

$$\alpha = \frac{n_e}{n_t} \tag{1.10}$$

Only a small fraction of the atoms in an electric discharge are ionized, typically 1 in 10^5–10^6, which at a gas pressure of 133 Pa (1 Torr) corresponds to a particle density of 10^{22} m^{-3} and an electron density of 10^{16} m^{-3}. The variation of the degree of ionization with temperature is illustrated in Figure 1.6.

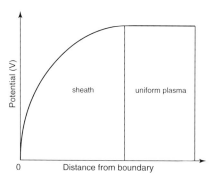

Figure 1.5 Illustration of the voltage distribution in a sheath at a boundary.

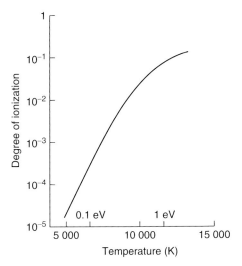

Figure 1.6 Characteristic variation of the degree of ionization of an atomic gas at atmospheric pressure.

1.4
Plasma Resonance

Resonance occurs in a plasma at high frequencies at different frequencies corresponding to the electron and ion frequencies. The electron frequency is independent of the mass of the atom or molecule and is sometimes referred to as the *plasma frequency*. The resonant frequency defines one of the characteristic properties of plasmas at high frequency, namely the cut-off frequency at which an electromagnetic wave will not be transmitted through a plasma. The conditions for this will be dealt with in Chapter 3 but can be considered in terms of the Debye radius (see Section 1.3.5). The Debye radius is the distance that an electron has to move to screen the plasma from radiation. If the period of the frequency of the radiation is such that it is shorter than the average time for an electron to move through the Debye radius, the energy will be transmitted since the electrons in the limit will not have time to absorb energy, whereas if it is longer energy will be absorbed. This accounts for the variation of the transmission of electromagnetic waves with frequency in the ionosphere; however, in the case of technological plasmas the distances involved are short and absorption within a narrow depth is normally required.

1.5
The Defining Characteristics of a Plasma

The high temperature and energy density of plasmas are illustrated in Figure 1.7, which also shows different methods of energizing plasmas. The defining

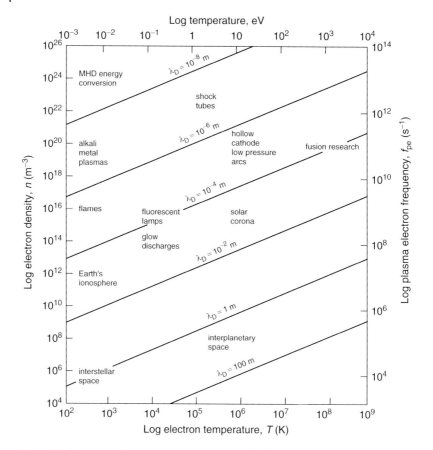

Figure 1.7 Variation of electron number density with electron energy at different values of Debye radius for different plasma processes.

characteristics of a plasma are shown in Figure 1.7, together with some characteristic ranges of plasma applications. The plasma conditions can be determined if any two parameters are known.

In addition to charge equilibrium, some additional defining characteristics of a plasma are as follows:

1) Sufficient charged particle density so that the plasma is not discontinuous.
2) Debye radius greater than the distance between particles and smaller than the characteristic length of the plasma, that is, the size of the vessel should be much greater than the Debye length.
3) The plasma electron frequency $\omega_{pe} = 2\pi f_{pe}$ should be much higher than the collision frequency if a signal is to propagate.

There are many examples of plasmas which do not meet the strict definition of charge equilibrium, such as in the electrode regions of electric discharges, and electron and ion beams, which are otherwise similar, or contribute to the production of a plasma; many plasma processes themselves are generated using electron and ion beams from the cathode or anode fall regions of a discharge.

References

1. Reutscher, A. (2001) Characteristics of low temperature plasmas under non-thermal conditions - a short summary, in *Low Temperature Plasma Physics* (eds R. Hippler, S. Pfau, M. Schmidt and K.H. Schoenbach), Wiley-VCH Verlag GmbH, Weinheim, pp. 29–54.
2. Jewett, J. and Serway, R.A. (2008) *Physics for Scientists and Engineers with Modern Physics*, 7th edn, Thomson Higher Education, Belmont, CA.
3. Das, A.C. (2004) *Space Plasmas: an Introduction*, Narosa, New Delhi.
4. Chen, F.F. (1984) *Introduction to Plasma Physics and Controlled Fusion*, 2nd edn, Plenum Press, New York.
5. Cambel, A.B. (1963) *Plasma Physics and Magneto Fluid Mechanics*, McGraw-Hill, New York.
6. Roth, J.R. (1995) *Industrial Plasma Engineering, Vol. 1, Principles*, Institute of Physics, Bristol.
7. von Engel, A. (1983) *Electric Plasmas and Their Uses*, Taylor & Francis, London.
8. von Engel, A. (1965) *Ionised Gases*, Oxford University Press, Oxford (reprinted 1994, American Institute of Physics, New York).

Further Reading

Brown, S.C. (1966) *Introduction to Electrical Discharges in Gases*, John Wiley & Sons, Inc., New York.
Browning, B. (2008) *Basic Plasma Physics*, Lulu Enterprises, Raleigh, NC.
Dendy, R. (ed.) (1993) *Plasma Physics: an Introductory Course*, Cambridge University Press, Cambridge.
Eliezer, Y. and Eliezer, S. (1989) *The Fourth State of Matter*, Hilger, Bristol; 2nd edn, 2001.
Howatson, A.M. (1976) *An Introduction to Gas Discharges*, 2nd edn, Pergamon Press, Oxford.
Suplee, C. (2009) *The Plasma Universe*, Cambridge University Press, Cambridge.

2
Elastic and Inelastic Collision Processes in Weakly Ionized Gases

2.1
Introduction

The behaviour of electrons with neutral particles in weakly ionized gases is similar to their behaviour in ideal gases and gives a useful insight into the behaviour of electrical plasmas. Elastic collisions (in which momentum is conserved) and inelastic collisions are the mechanisms by which energy is transferred from electrons in an electric field to neutral particles and form a plasma.

The simple ballistic equations (see Section 1.3.1) give an insight into the processes by which gain in energy occurs. At low pressures, where collisions are infrequent and the particle number density n is low, the behaviour of individual particles can be used as a basis to model processes. At high pressures, where the particle number density n and collision frequency v are large and the particle separation is small, the individual velocity is replaced by statistically averaged values.

The energy gained in elastic collisions over the mean free path λ (separation between particles) is transferred by inelastic quantum processes to potential energy in the form of excitation, ionization or dissociation and released by recombination. Finally the potential energy in the inelastic processes is transferred to the process.

Values of the mean free path, mean velocity, collision frequency and diameter are listed for different gases at room temperature and atmospheric pressure in Table 2.1 and are useful as a reference point at normal ambient conditions. Equivalent values for charged particles are less easy to determine since they are affected by interaction of their own and external electric fields.

2.2
The Drift Velocity

Electrons have a much higher drift velocity than ions, although in a plasma both are normally present in approximately equal numbers, maintaining charge equilibrium, and the high electrical conductivity of a plasma is normally due primarily to electrons [2].

Introduction to Plasma Technology: Science, Engineering and Applications. John Harry
Copyright © 2010 WILEY-VCH Verlag GmbH & Co. KGaA, Weinheim
ISBN: 978-3-527-32763-8

Table 2.1 Values of the mean free path λ_g of some gas molecules, their mean velocity $\bar{v}$ and their collision frequency v_c calculated from the kinetic theory of gases at $T = 288$ K and $p = 760$ Torr.

Gas	Molecular weight	λ_g (10^{-8} m)	$\bar{v}$ (m s^{-1})	v_c (10^9 s^{-1})	Diameter (10^{-10} m)
H$_2$	2.016	11.77	1740	14.8	2.74
He	4.002	18.62	1230	6.6	218
H$_2$O	18.000	4.18	580	13.9	4.60
Ne	20.180	13.22	550	4.2	2.59
N$_2$	28.020	6.28	467	7.4	3.75
O$_2$	32.000	6.79	437	6.4	3.61
Ar	39.940	6.66	391	5.9	3.64
CO$_2$	44.000	4.19	372	8.8	4.59
Kr	82.900	5.12	271	5.3	4.16
Xe	130.200	3.76	217	5.8	4.85

From Ref. [1].

The force on an electron of charge e and mass m_e in an electric field E is $eE = m_e a$, where a is the acceleration of the electron. The velocity u after a time t of a particle starting from rest is $u = at$, and the distance travelled is $ut/2$.

The electron velocity u_d constitutes a current $u_d e$. The current can be written in terms of the average drift velocity of electrons u_d in the direction of the electric field, so that

$$I = u_d n_e e \tag{2.1}$$

where n is the number of electrons.

The mean free path is the distance travelled by a particle before collision. Calculated values of the mean free paths of gas molecules, their mean velocities and their collision frequencies are shown in Table 2.1. The mean free path, collision frequency and their collision frequency scale fairly closely with gas pressure.

The behaviour of charged particles is less easy to determine by calculation as it is influenced by the proximity of other charged particles and electric fields and is more usefully found by measurement. The electron drift velocity may be more than 10^4 m s^{-1} at pressures below about 100 Pa (0.752 Torr) [2]; relatively few collisions occur at this pressure and the mean free path is governed by the electrode separation or vessel dimensions. At higher pressures, collisions with heavy particles reduce the time of acceleration in the electric field and the drift velocity decreases before collision, and at about 1 kPa (7.52 Torr) the thermal velocity is typically more than 100 times the drift velocity.

2.2.1
Electrical Conductivity

Electrical conductivity σ (Sm^{-1}) is defined as

$$\sigma = \frac{J}{E} \tag{2.2}$$

where J is the current density vector in Am^{-2} and E is the electric field strength in V m^{-1}. The resistance of a conductor $R = \rho l/A$, where ρ is the resistivity, l is the pathlength and A is the area of cross-section of the conductor perpendicular to the direction of current flow.

The electrical conductivity in terms of the number density of electrons n_e and the drift velocity using Eq. (2.2) becomes

$$\sigma = en_e \frac{u_d}{E} \tag{2.3}$$

The relatively high electrical conductivity of high-pressure plasmas is accounted for by the very large number of free electrons; for example, in air at atmospheric pressure with 1% ionization, using Loschmidt's number ($N_L = 2.69 \times 10^{25}$ particles m^{-3}) the number of electrons available is about 10^{20} m^{-3}, so that even a velocity of a few metres per second will correspond to a current density of several amperes per square metre.

2.2.2
Mobility

The mobility of an electron or ion [2, 3] is defined as

$$\mu = \frac{u_d}{E} \tag{2.4}$$

The electrical conductivity can be written in terms of the electron mobility μ_e as

$$\sigma = en_e \mu_e \tag{2.5}$$

where $\mu_e = u_d/E$ m^2 V^{-1} s^{-1} is a measure of the variation of the drift velocity with the applied electric field.

When the electric field is low (that is, the gain in energy from the applied electric field E is much less than the average change due to the thermal energy) and the pressure is below about 133 Pa (1 Torr), where the number of collisions is small, $\mu_e \propto E/p$, the mobility varies approximately linearly with the particle drift velocity and electric field strength (see Section 3.4); the mobility is of less value in describing the conduction process at higher gas pressures and higher electric fields where the effect of increased collisions with neutral particles and the effect of the electric field limit the gain in drift velocity $\mu_e \propto (E/p)^{1/2}$ [3]. The mobility and collision frequency vary widely with different gases and gas pressures but have been measured for different gases over a very wide range of conditions [1, 4].

2.2.3
Thermal Velocity

The thermal velocity of atoms and ions in gases is due to thermal diffusion along the temperature gradient and can be readily determined from the gas equation and their kinetic energy using the Boltzmann constant. The thermal velocity of electrons is considerably greater than that of ions or neutral particles due to their lower mass (ratio of mass of proton to mass of electron = 1833). Collisions between particles result in random motion being superimposed on the thermal velocity (Figure 2.1).

The thermal velocity u_r of gases at atmospheric pressure and 20 °C is of the order of 5×10^3 m s^{-1} and collisions with atoms and molecules reduce the drift velocity and are the cause of their high electrical resistivity ($>10^9$ Ω m) before breakdown. At low pressures, the drift velocity in an electric field may be much higher than the diffusion velocity. After breakdown, the DC electrical resistivity of plasmas is of the order of 10^{-9} Ω m^{-1}, comparable to the electrical resistivity of metals.

2.2.4
Collision Frequency

The collision frequency and collision cross-section are two important parameters that determine the energy density in a plasma and its behaviour over a very wide range of parameters. The collision frequency is difficult to analyse except at

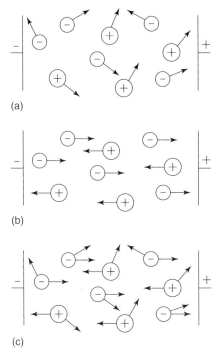

Figure 2.1 Motion of charged particles: (a) thermal velocity; (b) drift velocity component due to electric field; (c) combined effect of thermal velocity and drift velocity.

very low pressures. At high gas pressures, where the collision frequency is high, experimentally determined results for the collision cross-section are available for a very wide range of conditions [4].

Collision frequencies between electrons and atoms are normally high and account for the high energy densities obtained in plasmas. For example, the collision frequency of an electron with a neutral particle of an electron in neon at 133 Pa (1 Torr) is about $3 \times 10^8 \text{ s}^{-1}$.

If the collision with neutral particles is the main energy transfer process, the drift velocity is governed by the average time between collisions. If an electron collides after a time τ_{ce}, the final velocity is

$$u_d = \frac{eE}{m_e} \tau_{ce} \tag{2.6}$$

and the time between collisions $\tau_{ce} = \lambda_{ce}/u_d$, where u_d is the average velocity.

The mobility can be written in terms of the collision frequency ν_{ce} since $u_d = ft$ and $m_e f = eE$:

$$\mu_e = \frac{e}{m_e \nu_{ce}} \tag{2.7}$$

and the electrical conductivity is given by

$$\sigma = n_e e \left(\frac{e}{m_e \nu_{ce}} \right) \tag{2.8}$$

The electrical conductivity determined in this way is useful in describing DC plasmas. If the frequency of an AC supply is much lower than the collision frequency, the behaviour will be similar to that of a DC plasma, but if it is larger the behaviour is more complex (Chapter 3).

2.2.5
Collision Cross-section

The collision cross-section σ_c of a particle is the effective area of that particle over which a collision between two particles can occur. If one of the colliding particles is an electron, with an effective radius much smaller than that of any atom or molecule, then the collision cross-section is the cross-section of the atom or molecule involved in the collision with the electron, that is, $\sigma_c = \pi r^2$. For two particles of radii r_1 and r_2 the collision cross-section becomes $\pi (r_1 + r_2)^2$ and if they are of equal size collision can occur over an area $\sigma_c = 4\pi r^2$ (Figure 2.2).

Over the path travelled, the particle will collide as many times as the number of other particles contained within the cylindrical volume in a random walk.

The number of collisions by a particle moving at a velocity u can be determined using the collision cross-section, since a particle travelling at a velocity u for a time t sweeps out a cylindrical volume V_{vol} equal to $\sigma_c ut$ (Figure 2.3). If the number density of the gas is n_g, then the number of particles within the cylindrical volume is $n_c ut \pi r^2$, and therefore the number of collisions is given by $n_c \sigma_c ut$.

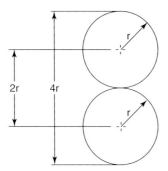

Figure 2.2 Collision cross-section for particles of equal size.

The mean free path is the total distance travelled in a time t divided by the number of collisions:

$$\lambda = \frac{ut}{n_c \sigma ut} = \frac{1}{n_c \sigma_c} \qquad (2.9)$$

The time between collisions is the velocity divided by the mean free path. It follows that the collision frequency v is

$$v = \frac{u}{\lambda} = n_c \sigma_c u \qquad (2.10)$$

The mean free path of an electron is affected by interactions with other charged particles and also the effect of collisions (Coulomb interaction). The electromagnetic forces between ions and electrons are high compared with the forces associated with a change in momentum. Electrons can be coupled to slow-moving ions, deflected or slowed by ions or influence ions (ambipolar diffusion) within the Debye length, except at low pressure where the distances separating them may be large.

The drift velocity of ions and electrons may be much greater than the thermal velocity at low pressures where few collisions occur and the simple relationships applicable to heavy particles and field-free regions are inaccurate. Collision cross-sections will exist for different gases and types of collision, electron energy

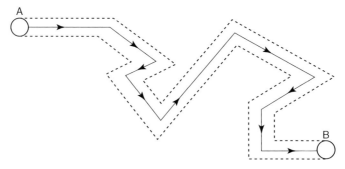

Figure 2.3 Collision path of a particle entering a field of slow-moving particles (no exchange in momentum).

and Coulomb interactions and the mean free paths of electrons and ions are often determined empirically [3].

2.3
Inelastic Collision Processes

Inelastic collision processes such as excitation and ionization are due to the change in energy between energy levels of an atom or molecule (Figure 2.4). Dissociation, by splitting up a molecule and storing potential energy, is also an inelastic process.

The ballistic equations that are applicable to elastic collisions do not apply to inelastic processes for which the energy is transferred in quanta. (Ballistic equations may still be used for electrons and ions involved in elastic processes.) Depending on the kinetic energy of the electron on impact with a neutral particle, an atom may be excited or ionized and a molecule may be dissociated, left in a vibrational state or ionized.

Energy is related to wavelength by $E = hf$, where E is a quantum of energy (eV), h is Planck's constant (6.626×10^{-34} J s) and the velocity of light $c_0 = f\lambda$. The radii of the different orbits are proportional to the square of the integers corresponding to the number of the orbit. The lowest energy corresponds to the outermost orbit, which when this is filled is known as the *ground state*. Electrons can gain energy by collision and transfer to another orbit further out, gaining potential energy and are excited or become free electrons and the particle is ionized. The energy required to free an electron from the outermost level is known as the *first ionization potential*. The energy transferred per transition of an atom or molecule is very small and is measured in electron volts (eV) (see Section 1.3.2).

The energy distribution as a function of wavelength and atomic transition is listed in Table 2.2.

The mean energy of ions in their own gas is approximately the same as that of the atoms or gas molecules because of their high mass and low drift velocity. The mean energy distribution is given by the Boltzmann equation (Eq. 2.6) so that we can write the ratio of the number of electrons in each state energies ϕ_1 and $\phi\zeta_2$ as

$$\frac{N_1}{N_2} = \exp\left[\frac{-(\zeta_2 - \zeta_1)}{k_B T}\right] \qquad (2.11)$$

A very large number of possible energy transitions exist between different levels and for different gases.

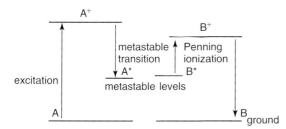

Figure 2.4 Illustrating energy levels associated with excitation, metastable and Penning transitions.

Table 2.2 Radiative processes and their characteristic wavelength, frequency and energies.

Frequency band	Wavelength range (m)	Frequency (Hz)	Energy (eV)	Type of process
Near-infrared	10^{-6}	3×10^{14}	1.2	Vibrational and electronic excitation
Visible light	5×10^{-7}	0.6×10^{15}	2.4	Bond breaking
Near-ultraviolet	3×10^{-7}	1×10^{15}	4.1	Outer shell electron liberation
Far-ultraviolet	1.5×10^{-7}	2×10^{15}	8.3	Middle shell electron liberation
Long X-ray	3×10^{-8}	1×10^{16}	4.1×10	Inner electron
Short X-ray	10^{-8}	3×10^{16}	1.2×10^{2}	Liberation
Gamma-ray	10^{-12}	3×10^{20}	1.2×10^{6}	Nuclear processes

2.3.1
Excitation

Excitation is the transition of an electron from the orbit of an atom or molecule to an outer orbit, usually by collision with a free electron. Normally it is the outer orbit since this requires least energy. Release of an electron requires an energy above about 0.01 eV. Excitation transitions occur up to energy levels of about 10 eV when excitation takes place and have a lifetime of about 10^{-9} s [5]. Excitation of an atom can only occur from an electron transition; molecules may also be energized by vibrational and rotational transitions. The process of electronic excitation of an atom is written as

$$e + A \rightarrow A^* + e$$
$$A^* \rightarrow A + hf \quad (2.12)$$

The recombination is accompanied by emission of energy, often as light at a wavelength corresponding to the excitation energy. The process for a molecule is more complex as there are more orbital transitions and several possible emission wavelengths.

The electron energy in a cold plasma is relatively low compared with the ionization potential, and only a relatively few electrons in the tail of the energy distribution have sufficient energy to cause ionization. Excitation acts as a short-term store of energy as the energy is continually transferred very rapidly between excited particles without losses to the surrounding medium. In this way, the low energy from electron collisions is amplified by many collisions and storage over a short period where the effect of collisions is additive to a level at which ionization can occur.

2.3.1.1 Metastable Processes

A transition to the ground level is not permitted in some atoms and molecules such as the noble gases and mercury, which have metastable energy states. The noble gases have metastable states in the range 10–20 eV (Table 2.3).

Table 2.3 Metastable energies[a] of the noble gases.

Gas	Metastable energy (eV)
He	19.8
Ne	16.6
Ar	11.5
Kr	9.9
Xe	8.32

[a] See also Table 2.5.

Excitation transitions of metastable states have lifetimes of up to 10×10^{-3} s [5]. A metastable atom may collide with a ground (unexcited)-state atom, causing excitation:

$$A + e \rightarrow A^* + e \quad (2.13)$$

Since no electron is released, this is a resonant process and the ionization energy is reduced by $V_i - V^*$. A further collision with an electron will increase the energy of the atom and may cause ionization in a stepwise manner:

$$A^* \rightarrow A^+ + e, \ e + A^* \rightarrow e + e + A^+ \quad (2.14)$$

The metastable level has the effect of increasing the collision cross-section and enabling higher energy levels to be obtained from the initial coexcitation collision and subsequent collisions which may be sufficient to ionize the gas [4].

2.3.2
Ionization and Recombination Processes

When the kinetic energy of an electron in collision with an atom or molecule exceeds the ionization energy eV_i of an atom or molecule, excitation or ionization may occur and a positive ion and two slow-moving electrons result:

$$e + A \rightarrow 2e + A^+ \quad (2.15)$$

The new electron may in turn produce another electron by impact as it gains energy in the electric field resulting in a multiplication process leading to a sustained plasma.

Ionization generally occurs at energy levels higher than excitation levels, corresponding to frequencies greater than 2×10^{14} Hz (Table 2.4). The ionization energy increases with distance of the electron from the nucleus due to the shielding effect of the inner shells and the reduced distance from the nucleus. Typical values of ionization energies corresponding to the first and second levels of ionization normally encountered in electric discharges are between 10 and 30 eV. For example,

Table 2.4 Some ionization processes.

$e + A \rightarrow A^+ + 2e$	Ionization by electron–atom collision (>10 eV)
$e + A^* \rightarrow 2e + A^+$	Electron metastable ionization (eV)
$e + AB \rightarrow AB^-$	Ionization by electron attachment
$A^* + B \rightarrow A + e + B^+$	Metastable neutral ionization collision (Penning ionization) Energy of B < excitation energy of A^*
$A^+ + B \rightarrow A + B^+$	Charge exchange
$A^+ + B \rightarrow A^+ + B^+ + e$	Ionization

the first ionization potential of argon is 15.7 eV, helium 24.6 eV, nitrogen 4.54 eV and oxygen 13.6 eV. Table 2.4 gives examples of ionization processes.

2.3.2.1 Charge Transfer

Charge transfer can take place in a collision between an atom and an ion in which an electron is removed from the atom. The neutral atom ionizes and excites an upper state of an ion formed by the collision with the atom:

$$A^* + B \rightarrow A + (B^+)^* \pm \Delta E \quad (2.16)$$

If the ion and atom are the same species, no change in overall kinetic energy occurs and no energy is released as it is a resonant process.

2.3.2.2 Dissociation

Dissociation of gases generally occurs from around 1000 K up to about 4000 K, at which dissociation is complete. Typical bond dissociation energies are between about 3 and 7 eV, but can be as high as 9.8 eV/(N_2) and 11 eV (CO), sufficient for excitation or ionization. Dissociative ionization can may also occur by ion pair formation [5]:

$$e + A_2 \rightarrow A^+ + A^- + e \quad \text{and} \quad e + AB \rightarrow A^+ + B^- + e \quad (2.17)$$

2.3.2.3 Negative Ionization

Negative ions may be formed by dissociative attachment and form negative ions with large collision cross-sections at low energies (<1 eV) in electronegative gases such as the halogens and oxygen. These have completed inner shells, form a weak bond with free electrons and form electronegative ions:

$$e + AB \rightarrow A + B^- \quad (2.18)$$

2.3.2.4 Recombination

Recombination is the process by which the plasma loses the potential energy it has gained by recombination of electrons and ions in a radiative recombination process:

$$e + A^+ \rightarrow A + hf \tag{2.19}$$

or by dissociative recombination:

$$AB^+ + e \rightleftharpoons AB^* + B \tag{2.20}$$

$$e + A_2^+ \rightarrow 2A \tag{2.21}$$

The mean free path of ions in a discharge at atmospheric pressure is less than 10^{-7} m and in a plasma reactor the probability of an ion encountering a solid surface before a collision takes place is very small, so that volume recombination dominates. Recombination between electrons and ions occurs only at the walls of the containing vessel at low pressures where the mean free path is large compared with the vessel. Boundary collisions are an important part of most plasma processes. An example is an ion A^+ that collides with a boundary and is neutralized, resulting in the emission of an electron by secondary emission from the surface:

$$A^+ \rightarrow A + e \tag{2.22}$$

The subsequent recombination between negative ions and positive ions has a higher probability than recombination between electrons and positive ions because in the former case the two ions possess comparable kinetic energy.

2.3.2.5 Metastable Ionization

The effect of metastable states or Penning ionization (see Chapter 3) [6] on the number of collisions increases the likelihood of a transition and has a similar effect to increasing the collision cross-section, which is a useful way of describing the processes.

The long lifetime of metastable excitation assists the build-up of ionization by subsequent collisions but is not normally the main ionization process. The transitions depend on the gas pressure, gas, vapour and impurities, which even in very small quantities may have a major affect on the transitions by providing intermediate and metastable energy levels, used to advantage in lamps and lasers.

Metastable ionization enables a nonresonant energy interchange to take place by charge transfer in a gas mixture in which the second constituent has a lower ionization potential (Penning ionization). Penning ionization occurs when an excited neutral atom collides and ionizes another neutral atom:

$$A^* + B \rightarrow A + B^+ + e \text{ or dissociation or } A^* + B_2 \rightarrow A + 2B \tag{2.23}$$

Noble gases such as helium and neon may be added to a gas to use the Penning affect to enhance the excitation or ionization processes since, although they have a high ionization energy, they have long metastable states enabling resonant charge interchange:

$$A^* + B \rightarrow A + B^* \tag{2.24}$$

or even ionization:

$$A^* + B \rightarrow A + B^+ + e \tag{2.25}$$

Table 2.5 Energies characteristic of some atoms and molecules [4].

Atom or molecule	Ionization potential (eV)	Electron affinity (eV)	Metastable energy level[a] (eV)	Lowest excitation energy (eV)	Dissociation energy (eV)
H	13.6	0.75	–	10.2	–
He	24.6		19.8	21.2	–
O	13.6	1.5	1.97	9.2	–
F	17.4	3.5	–	12.7	–
Cl	13.0	3.6	–	8.9	–
H_2	15.6		–	11.5	4.5
O_2	12.5	0.45	1.2	7.9	5.1
Cl_2	13.2	2.5	–	–	2.5
C_6H_6	9.6	–	–	–	–

[a] See also Table 2.3.

Metastable ionization enables resonant interchange to take place by charge transfer in a gas mixture in which the second constituent has a lower ionization potential (Penning ionization).

The ionization potential of a gas is a measure of the energy required to ionize the gas and the potential energy available in a plasma process. Table 2.5 shows the ionization potential, the minimum excitation energy and dissociation energy for a number of gases. The lowest excitation level indicates the difficulty of excitation and where it is close to the ionization potential. The small dissociation values indicate the likelihood of atom and molecules coexisting.

The noble gases with metastable states and metal vapours and atoms with high atomic numbers are easier to ionize than atoms with low atomic numbers such as helium or argon or polyatomic gases.

At low pressure, the reduced number of collisions results in higher particle energies and ionization becomes easier until at even lower pressures the reduced number of collisions results in higher voltages.

References

1. McDaniel, E.W. (1964) in Collision phenomena in ionised gas, in *Fundamentals of Gaseous Ionisation and Plasma Electronic* (ed. E. Nasser), John Wiley & Sons, Inc., New York.
2. von Engel, A. (1965) *Ionised Gases*, Oxford University Press, Oxford (reprinted 1994, American Institute of Physics, New York).
3. von Engel, A. (1983) *Electric Plasmas and their Uses*, Taylor & Francis, London.
4. Brown, S.C. (1959) *Basic Data of Plasma Physics: the Fundamental Data on Electric Discharges in Gases*, MIT Press, Cambridge, MA (reprinted 1997, Classics in Vacuum Science and Technology, Springer, New York).

5. Roth, J.R. (1995) *Industrial Plasma Engineering, Vol. 1, Principles*, Institute of Physics, Bristol.
6. Graham, W.G. (2007) Introduction – the potential of plasma technology in the textile industry, in *Plasma Technology for Textiles* (ed. R. Shishoo), Woodhead Publishing, Abington, Cambridge, pp. 1–24.

3
The Interaction of Electromagnetic Fields with Plasmas

3.1
Introduction

This chapter considers the many ways in which electrical energy can be coupled to a plasma. These include DC and AC, using electrodes and electrodeless coupling at high frequencies, resonant coupling processes and magnetized plasmas and the propagation and absorption of electromagnetic waves in plasmas.

3.2
The Behaviour of Plasmas at DC and Low Frequencies in the Near Field

The properties of an electromagnetic wave vary in the near and far field as a function of its wavelength; for example, at 50 kHz the wavelength is 6 km and at 5 GHz it is 60 mm. In the near field, although the coupling process is complex, it can usually be considered in simple terms since the near field dominates. Both electric and magnetic fields are high in the near field and are used in inductively coupled plasma (ICPs) and capacitively coupled plasmas (CCPs), and it is often sufficient to consider the coupling process in terms of simple processes such as resistors capacitors and transformers.

At DC and low frequencies, electrons and ions in plasmas normally behave as coupled charges, that is, positive ions tend to restrain the movement of electrons. At AC frequencies, as the frequency is increased the electrons and ions respond differently because of their different masses. The conduction of DC in a plasma at around atmospheric pressure is by both electrons and ions and the electrons and ions behave so as to cancel their individual charges in a similar way to current flow in a solid conductor.

At low pressures, as the spacing between particles increases and the number of collisions decreases, the coupling between the particles is reduced, the electric field affects the electrons and ions differently and the behaviour is described as particle behaviour. In the extreme, at high electric field strengths, moderate gas pressures

Introduction to Plasma Technology: Science, Engineering and Applications. John Harry
Copyright © 2010 WILEY-VCH Verlag GmbH & Co. KGaA, Weinheim
ISBN: 978-3-527-32763-8

and correspondingly high collision frequencies, the behaviour is kinetic whereas at low pressures it is described as free electron (or ion), for example in electron and ion beams.

At atmospheric pressure, there are wide differences between the velocities and energies of individual particles and local charge equilibrium is maintained. Energy transferred through collisions may be low but the collision frequency is very high ($>10^{34}$ m^3 s^{-1}). The overall behaviour is that of coupled charges down to low pressures of about 1 kPa (7.52 Torr) and electron number densities are below about 10% where the particles, electrons and ions have a high collision frequency and remain in close proximity to each other.

Over this region, the plasma behaviour may be represented statistically; energy acquired between collisions is small and the collision frequency is high. The current is limited by the series impedance of the supply (see Chapter 9). The voltage can be controlled by varying the gas composition and by constricting or diffusing the plasma. The power input is unlimited and the problem is usually to obtain sufficient energy density and limit the power from the supply by electrical stabilization.

AC conduction occurs in a similar way to DC conduction, although the inertia of the ions becomes apparent at frequencies of about 1 kHz (see Section 3.2.1.1). Below the electron plasma resonant frequency $\omega < \omega_{pe}$, the particles behave in the coupled mode. Over the collective region, the behaviour in an electromagnetic field is similar to the current in a solid but flexible conductor and the mechanism of energy absorption from an electromagnetic field is analogous to resistive losses at DC and low frequencies. As the frequency is increased, a frequency is reached at which the ions resonate and as the frequency is increased further electrons resonate. The frequency of resonance of ions varies with the molecular weight of the ion but the resonant frequency of an electron is not affected by the molecular weight and is often referred to as the *plasma frequency*. As the electron density increases, the plasma frequency and the frequency at which collective behaviour occurs increases.

Above the electron resonant frequency, at high electron number densities the particle movement is small, the number of collisions is high and the behaviour is collective; that is, they behave in a similar way to the electrons and ions in a conductor and particles are not affected individually.

At low pressures, the number of collisions is small and it is difficult to obtain a high enough power and energy intensity to carry out useful processes. Below pressures of about 1 Pa (7.52 $\times$ 10^{-3} Torr), the particles no longer behave as if they were coupled. Electrical conduction processes follow simple ballistic equations where few collisions occur, such as in electron and ion beams.

At very low pressures and at high frequencies, damping by collision of charged particles is reduced, simple harmonic motion and resonance occur and the current may be carried by oscillation of charge. The reactor walls may acquire a collisionless charge of alternating polarity by a displacement current, with few if any collisions with particles or walls and electrodes.

3.2.1
Charged Particles in Electromagnetic Fields

Electromagnetic fields enable energy to be coupled to a plasma. The force acting on a charge q moving in an electric field with the magnetic field is given by the vector addition of the electric and magnetic forces:

$$\overline{F} = q[\overline{E} + (\overline{u} \times \overline{B})] \tag{3.1}$$

A charge in a DC electric field between two plane electrodes is shown in Figure 3.1a (the charge is an electron so it will move in the opposite direction to the conventional flow of current) [1].

If the magnetic field is parallel to the electric field, it has no effect on the electron and simplifies the algebra, and Eq. (3.1) simply becomes

$$F = eE \tag{3.2}$$

The velocity u of an electron after a time t is $u = ft$. This energy is normally transferred to electrons and ions, and then to atoms, some of which absorb the energy in elastic and inelastic collisions and become excited or ionized, releasing more electrons.

A moving charge constitutes a flow of current qu (Figure 3.1b). When an electron moves in a magnetic field perpendicular to the direction of its travel, it results in a force on the charge:

$$F = Beu \tag{3.3}$$

If a stationary charge is in an electric field, it has an electrostatic force acting on it, and if it moves it behaves like a current and produces a magnetic field.

The ratio of the electric and magnetic forces is

$$\frac{Bue}{eE} = \frac{Bu}{E}$$

$$B = H\mu_0 \quad \text{and} \quad \frac{E}{H} = \sqrt{\frac{\mu_0}{\varepsilon_0}}$$

$$H\frac{\mu_0}{E}u = \mu_0 u \sqrt{\frac{\varepsilon_0}{\mu_0}} = u\sqrt{\mu_0 \varepsilon_0} = u/c_0 \tag{3.4}$$

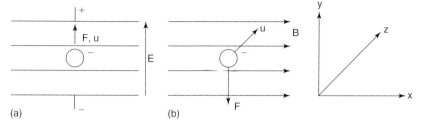

Figure 3.1 Forces acting on a charge moving in (a) an electric field and (b) a magnetic field.

As the drift velocity is usually very much less than the speed of light, the effect of the magnetic field of an electromagnetic wave can usually be ignored.

3.2.1.1 Behaviour of a Charged Particle in an Oscillating Electric Field

In the event of a head-on collision with a neutral particle, the total momentum of an electron is transferred to the particle [2] that is it starts from rest after a collision. If the average time between collisions between electrons and atoms is τ_{ce}, the average final velocity before collision is

$$u_{av} = \frac{eE}{m_e}\tau_{ce} \tag{3.5}$$

and the electric field strength is

$$E = \frac{u_{av} m_e}{e} v_{ce} \tag{3.6}$$

where v_{ce} is the collision frequency $= 1/\tau_{ce}$.

An electron in an AC electric field with peak amplitude $\hat{E}$ and angular frequency ω_s has a force acting on it $e\hat{E}\cos\omega_s t$ [2]. (We shall use $\omega(2\pi f)$ as the angular frequency, as this simplifies the algebra.) If no collisions occur, the equations of motion (see Chapter 2) become

$$m_e a = e\hat{E}\cos\omega_s t \tag{3.7}$$

the velocity

$$u = \frac{e\hat{E}}{m_e \omega_s}\sin\omega_s t \tag{3.8}$$

and the displacement

$$s = -\frac{e\hat{E}}{m_e \omega_s^2}\cos\omega_s t \tag{3.9}$$

Figure 3.2 shows an electron and ion in an electric field oscillating between the electrodes along the conducting pathlength. As the supply frequency is increased, a frequency is reached when a charged particle starting from rest is unable to traverse the path (electrode gap or the equivalent pathlength of an electrodeless plasma). The particle is trapped in the gap until it relaxes and loses its charge ($\sim 10^{-10}$ s) for an electron, but rather longer for an ion.

The drift velocity is very much smaller for an ion so that the maximum frequency at which the electron is able to cross the gap is higher for an electron than an ion. The persistence of ionization in the gap depends on the relaxation time, which is generally longer than that of an electron due to the greater mechanical and thermal inertia of an ion.

At moderate gas pressures, the drift velocity is a function of both the supply and collision frequencies. At high frequencies and at low pressures, the electron oscillates in the electric field without collision, whereas the ion, because of its higher inertia, remains approximately stationary.

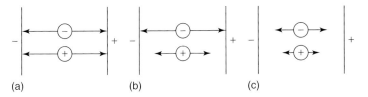

Figure 3.2 Oscillation of a charged particle in an electric field along its pathlength at different frequencies. (a) DC; (b) period of supply longer than the time for the ion to traverse the pathlength, but shorter than for an electron; (c) period of supply shorter than for the electron or ion to traverse the pathlength.

The effect of the thermal inertia of the ionized gas in the gap increases with frequency and the plasma becomes easier to ignite and maintain as the frequency increases due to the persistence of ionization.

If an electron is accelerated over a distance d and collides with a neutral particle and transfers all its momentum the energy transfer:

$$E = eEd \tag{3.10}$$

where E is the r.m.s. value of the electric field. The mean power per unit volume is

$$P = eEv_e n_e \tag{3.11}$$

When collisions occur at intervals greater than π/ω_s (the time available for acceleration), using Eq. (3.5) the maximum velocity of a charged particle is $eE/m_e\omega_s$ and the mean power per unit volume is

$$P = \frac{e^2 E^2 n_e}{2m_e \omega_s^2} \tag{3.12}$$

The E^2 term indicates that the electron gains energy when moving with or against the direction of the field [2]. The ω term shows that the energy decreases with supply frequency as the time of acceleration decreases before the electric field direction is reversed. From Eq. (3.12), it can be seen that when an elastic collision occurs in a radiofrequency (RF) plasma between an electron and an atom, if the direction of the electric field is approximately in the same direction of the velocity of the electron after collision the electron will continue to gain energy and it is this that may explain the ability of RF sources to ionize and maintain plasmas [2].

The derivation of the equation for the average RF power in a volume of gas is lengthy [3], but we may write the following for an electron rotating in the same plane as the rotation of the magnetic field (right polarized so that the electrons are subject to a reduced frequency) [4]:

$$P_{av} = \frac{n_e e^2 E_0^2}{2m_e v_{ce}} \left(\frac{v_{ce}^2}{v_{ce}^2 + \omega_s^2} \right) \tag{3.13}$$

where ω_s is the supply frequency. The term $v_{ce}^2/(v_{ce}^2 + \omega_s^2)$ is a measure of the effectiveness of the energy transfer. When the collision frequency is much higher than the supply frequency, $v_{ce} \gg \omega_s$, the transferred power is not significantly affected by the supply frequency. (The apparent inconsistency in collision frequency v_{ce} and the angular frequency of the supply ω_s in the denominator arises from the derivation of the constants for the condition which assumes the entire momentum in the direction of the electric field is given up.)

At 10^3 Pa (7.52 Torr), typical of glow discharge applications, the degree of ionization is less than 10^{-4} and the elastic collision frequency between electrons and neutrals is normally much higher than the supply frequency. The electrons experience many collisions during an RF cycle, and the gain in velocity and kinetic energy is small. The gain in kinetic energy is made up of frequent collisions between electrons, ions and neutrals.

As the pressure is decreased, the proportion of electron–ion collisions increases and, at pressures below about 1 Pa (7.52×10^{-3} Torr), the degree of ionization may be above 10^{-2} and electron–ion collisions predominate. As the pressure decreases further, collisions decrease and for $v_{ea}/\omega_s < 1$ at very high frequencies the energy input is enhanced by oscillation of the charged particle at the plasma frequency and collisionless absorption, for example by resonance at the plasma frequency, becomes the main mechanism of absorption of power in the plasma.

3.2.1.2 Plasma Frequency

Plasmas respond to disturbances in a similar way to mechanical systems.

Figure 3.3a shows a volume of plasma containing an equal number of positive ions and electrons in charge equilibrium. If a sudden charge perturbation of the electric field occurs in a plasma, the charged particles will be accelerated under the influence of the change in the electric field towards the electrode of opposite polarity and cause an equal and opposite field. Figure 3.3b shows the effect of a perturbation. The electrons move back rapidly to restore equilibrium and overshoot

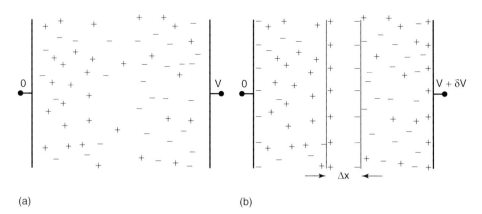

Figure 3.3 Imbalance of charge density in a plasma due to a perturbation (a) charge distribution in equilibrium, (b) effect of disturbance.

the equilibrium position and oscillate; the heavier positive ions move more slowly so that a charge imbalance exists a condition for resonance can occur.

The force on an electron is

$$eE = -m_e \frac{d^2 x}{dt^2} \quad (3.14)$$

Poisson's equation for the variation of the electric field in one dimension:

$$\frac{dE}{dx} = \frac{n_e e}{\varepsilon_0} \quad (3.15)$$

gives

$$E = \frac{n_e e}{\varepsilon_0} x \quad (3.16)$$

so that

$$\frac{d^2 x}{dt^2} = -\left(\frac{n_e e^2}{m_e \varepsilon_0}\right) x \quad (3.17)$$

which is the equation for second harmonic motion for which the angular frequency (rad s^{-1}) is

$$\omega_e = \left(\frac{n_e e^2}{m_e \varepsilon_0}\right)^{\frac{1}{2}} \quad (3.18)$$

The ion RF ω_i for the same charge density is

$$\omega_i = \sqrt{\frac{m_e}{m_i}} \omega_e \quad (3.19)$$

Ions behave in a similar way; however, the resonant frequency ω_i is reduced by the square root of their mass, $m_i^{1/2}$, and for a proton is 43 times less than the electron frequency.

When electrons oscillate at the same frequency as the electromagnetic wave, they resonate and interact and couple energy from the wave by Landau damping (as opposed to coupling by collisions).

Loss mechanisms of collision and resonance Landau damping [5] are two distinct mechanisms. If the supply frequency is much greater than the collision frequency, $\omega \tau_c \gg 1$, there will be several oscillations before a collision occurs.

The resonant wave is a longitudinal wave, that is, the E field is in the direction of travel of the electron rather than perpendicular to it as in the case of a transverse wave, as usually occurs with a typical electromagnetic wave.

3.2.1.3 The Debye Radius

A plasma can be visualized as a sea of equal numbers of electrons and positive ions. If the charged particles comprising the plasma are close enough to each other so that each particle also influences a number of adjacent particles, electrons are able to respond to screen out any disturbance below the electron resonant frequency ω_{pe} (ions can also, but the electron is more mobile). The effect is illustrated in Figure 3.4.

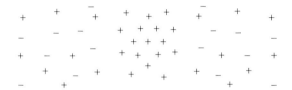
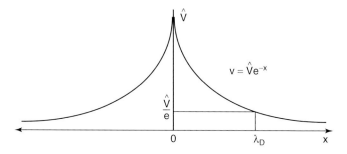

Figure 3.4 Illustration of Debye radius.

Charge imbalance often occurs in electric discharges and plasmas, at electrodes or probes, insulating surfaces or the walls of the chamber containing the plasma. The potential difference caused by the imbalance of charge, either at charged surfaces or the reduced number density of the faster electrons, results in the production of a local electric field gradient and charged particles of the opposite polarity form a sheath which screens the local electric field.

The electrostatic force between two point charges q_1 and q_2 separated by a distance r (see Section 1.3.4) is

$$\frac{q_1 q_2}{4\pi r^2 \varepsilon_0} \tag{3.20}$$

The electric field at distance r due to a single point charge at a distance r

$$E = \frac{e}{4\pi r^2 \varepsilon_0} \tag{3.21}$$

and potential

$$V = \int_\infty^r \frac{q}{4\pi r^2 \varepsilon_0} dr = -\frac{q}{4\pi r \varepsilon_0} \tag{3.22}$$

which attracts charges of opposite polarity that screen it within a sphere of radius r.

The derivation of the Debye shielding radius λ_D at which V is $1/e$ of its maximum value is lengthy [5], but

$$\lambda_D = \left(\frac{\varepsilon_0 k T_e}{n_e e^2} \right)^{\frac{1}{2}} \tag{3.23}$$

substituting for the fixed values for an electron gives

$$\lambda_D = 6.93 \times 10^{-3} \left[\frac{T_e \text{ (K)}}{n_e}\right]^{\frac{1}{2}} = 743 \left(\frac{T_e \text{ (eV)}}{n_e}\right)^{\frac{1}{2}} \tag{3.24}$$

$$\lambda_D = 69.1 \left(\frac{T_e \text{ (K)}}{n_e}\right)^{\frac{1}{2}} = 7440 \left(\frac{T_e \text{ (eV)}}{n_e}\right)^{\frac{1}{2}}$$

For an electron with an electron temperature $T_e = 1$ eV and charge density $N_e = 10^{16}$ m^{-3}, the Debye radius $\lambda_D = 74.4 \times 10^{-6}$ m and the number of particles in a Debye sphere with a radius equal to the Debye radius is $N_D = 17.3 \times 10^3$. The Debye radius is related to the plasma frequency so that for an electron

$$\lambda_D \omega_e = \left(\frac{\varepsilon_0 k T_e}{n e^2}\right)^{\frac{1}{2}} \left(\frac{n e^2}{m_e \varepsilon_0}\right)^{\frac{1}{2}} = \left(\frac{k T_e}{m_e}\right)^{\frac{1}{2}} \approx u_{eav} \tag{3.25}$$

This indicates that the electrons can move over a distance of one Debye radius during the period of one plasma oscillation; that is, a plasma will maintain neutrality by shielding the perturbation at a frequency below ω_e, but will not shield at higher frequencies. If the boundary or an electrode carries current, the distribution of charge in the sheath is more complex as it also serves to accelerate electrons from the boundary surface and carries the current.

A condition for a plasma to exist is that the sheath depth is sufficient to reduce the change in potential by 1/e from the equilibrium value and the effect of the disturbance is less than a critical dimension such as the size of the vessel [6].

Many applications using electric discharges fall outside this definition, such as free electron and ion process regions close to boundaries and electrodes, but otherwise have similar characteristics to a plasma. For example, the cathode of an electric discharge has a positive voltage gradient sheath due to the absence of high-energy electrons close to the cathode, which produces a voltage gradient sufficient for the cathode to emit electrons. The anode acts as a sink with a negative voltage gradient in front of it due to the absence of positive ions in this region.

Sheaths are unavoidable in most technological plasmas. In some applications they can be used to increase the kinetic energy of charged particles so as to change the surface properties, such as during the manufacture of computer chips and plasma nitriding.

3.3
Behaviour of Charged Particles in Magnetic Fields (Magnetized Plasmas)

Magnetic fields do not normally enable energy to be coupled to a plasma or charged particles, but have useful properties in controlling and assisting the coupling of energy from an electric field.

An electron moving in a magnetic field has a component of velocity corresponding to the initial drift velocity u_d which is equal to the tangential velocity u_t and a kinetic energy $m u_d^2 / 2$.

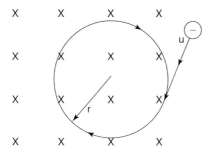

Figure 3.5 Behaviour of a charged particle entering a transverse magnetic field.

The locus of the particle is a circle if the particle enters the magnetic field in the plane perpendicular to the magnetic field flux lines. Figure 3.5 shows a charged particle (free electron or ion) moving at a velocity u in a collisionless medium. The moving electron constitutes a current eu so that for the forces to balance

$$F = euB = \frac{mu^2}{r} \tag{3.26}$$

The charge takes up a circular path at a constant radius known as the *gyro* or *Larmor radius*:

Figure 3.6 shows the variation of the gyro frequency of an electron with magnetic flux density entering a magnetic field. The gyro radius of an electron is typically measured in millimetres.

$$r = \frac{mu}{eB} \tag{3.27}$$

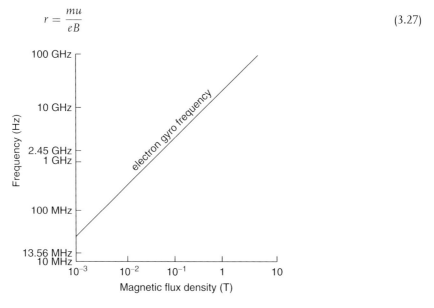

Figure 3.6 Variation of electron gyro frequency with magnetic flux density.

The time taken for one revolution is $2\pi r/u_t$, the angular velocity is

$$\omega_g = \frac{eB}{m} \tag{3.28}$$

and the gyro frequency is

$$f_g = \frac{eB}{2\pi m}$$

For a proton travelling at the same velocity the radius will be 1833 times larger and the frequency of rotation correspondingly less.

The orbit radius depends on the particle velocity entering the magnetic field and also its mass. For example, for a particle with an energy of 10 eV entering a magnetic field of 0.01 T, the electron velocity is 1.87×10^6 m s^{-1} and the radius is 1.064 mm, whereas for a proton the velocity is 4.38×10^4 m s^{-1} and the radius is 45.7 mm. Ion energies of 10–25 eV can be obtained, the corresponding energy for an electron being 5 eV.

If the electric field is perpendicular to the magnetic field, a charge enters the magnetic field at right-angles to the direction of the magnetic field and a drift velocity will be superimposed on the circular motion. The effect will be alternately to accelerate and decelerate the particle so that the particle traces a cycloidal path (Figure 3.7) and the angular velocity is refered to as the cyclotron velocity ω_c. The rotation of ions and electrons will be in the same direction if the axial velocities are in the opposite direction, and be in the opposite direction if the electrons and ions are travelling in the same direction.

The magnetic moment M of a charged particle in a magnetic field is an adiabatic invariant [6] for all except abrupt changes in magnetic field strength and is defined as the current I flowing in the gyro orbit multiplied by the area enclosed by the gyro orbit moment, $M = I\pi a^2$ where a is the area of the loop. The rotating electrons can be considered as a circulating current loop with a time of revolution $2\pi a/u_d$ and

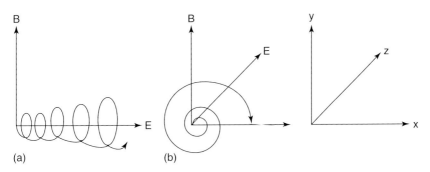

Figure 3.7 Motion of an electron entering crossed electric and magnetic fields on the axis of the electric field (a) side view, (b) end view.

an effective current $I = eu_d/2\pi a$, then

$$M = \left(\frac{eu_d}{2\pi a}\right)\pi a^2 = \frac{eu_d}{2a} \tag{3.29}$$

The radius of the loop $a = mu_d/eB$ so that

$$M = \left(\frac{eu_d}{2}\right)\left(\frac{mu_d}{eB}\right) = \left(\frac{mu_d^2}{2B}\right) \tag{3.30}$$

If a positive particle has a velocity driving it into a solenoid, the magnetic field increases and the radius of rotation decreases as it spirals in the axial direction towards the centre. Since the magnetic moment remains constant, the axial velocity decreases as B increases until the velocity along the magnetic field line is zero at a point known as the *mirror point*. The position of the mirror point depends on the axial component of velocity and the charged particle will be reflected in the opposite direction if the mirror point is before the centre of the solenoid where the magnetic field is highest. Two axial solenoids connected so that the direction of their magnetic fields oppose can act as a mirror or bottle trapping electrons.

The spiral path produced by crossed electric and magnetic fields (Figure 3.8) can be used to increase the pathlength of charged particles and the number of collisions by electrons with neutrals and ions.

Electrons and ions move along the line of flux produced by the magnet, but require energy from the electric field perpendicular to the magnetic field to cross lines of flux.

The electrons rotate around the lines of flux at the gyro frequency with a radius of typically 0.01–1 mm (correspondingly larger for ions), trapping them and further increasing their pathlength. The force due to the electric field eE accelerates the electrons in the direction of the electric field gradient and the electrons describe a spiral path of increasing radius as the flux density decreases, but decreasing velocity along the line of flux. The increase in the pathlength also increases the resistivity of a magnetized plasma.

The absorbed energy is a function of the collision frequency ν so that $\nu \ll 2\pi\omega_c$ to achieve high particle energies, that is, there needs to be several revolutions by an electron before a collision occurs.

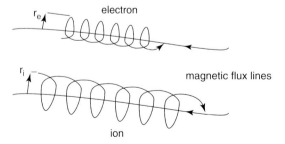

Figure 3.8 Motion of electrons and ions along magnetic field lines in a magnetron.

A magnetic field with its axis parallel with the path of a charged particle or a current has no effect on it, but if there is a radial component of velocity, as there almost inevitably will be because of collisions except at very low gas pressures, then the component of velocity or current perpendicular to the direction of the magnetic field will experience a force and rotate around the field line in a similar way to the behaviour in a magnetron (see Section 5.2.2). Any divergence of the magnetic field, for example at the ends of a bar magnet or solenoid, will also tend to converge or diverge any particles parallel to or on the axis of a magnet or solenoid.

3.4
Initiation of an Electrical Discharge or Plasma

Free electrons and ions produced by gamma-rays or photoionization can be accelerated by an electric field and a very small current flows. The electric field can be increased until all the free electrons are collected, that is, saturation occurs. If the electric field is increased further, acceleration of the electrons results in charge multiplication by ionizing collisions with neutral particles and the current becomes $I = I_0 e^{\alpha d}$, where α is the first Townsend coefficient [6]. If the loss coefficient corresponding to nonionizing collisions is γ, then when $\gamma(e^{\alpha d} - 1) > 1$ the rate of production of electrons exceeds the losses and the current increases rapidly and breakdown occurs. The equation holds for moderate pressures and electric fields.

The condition for the breakdown voltage can be written in the form of the probability of a collision which decreases as the gas pressure is reduced. The equation for the minimum breakdown voltage, Paschen's minimum sparking potential for initiating a discharge or plasma [7] becomes:

$$V_s = \frac{B(pd)}{C + \ln(pd)} \tag{3.31}$$

follows from Townsend's equation, where B and C are constants for different gases.

A characteristic curve for the sparking potential is shown in Figure 3.9. At pressures below the minimum sparking potential E/p is no longer constant and begins to increase, so that the pathlength necessary for breakdown increases as the pressure decreases.

3.5
Similarity Conditions

Consideration of Townsend's first coefficient also leads to the similarity principle, which is useful in predicting the effects of changes in plasma parameters [8].

Two discharges are similar if for the same voltage they carry the same current irrespective of their dimensions (Figure 3.10). For this to be true, the mean free path $\lambda \ll$ the pathlength d.

If the ionization rate is α, then for equal current

$$\alpha_1 d_1 = \alpha_2 d_2 \tag{3.32}$$

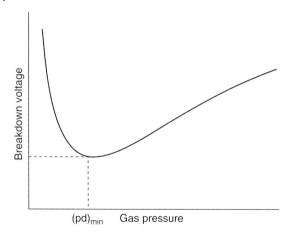

Figure 3.9 Illustration of minimum sparking potential.

and if the number of collisions is proportional to the gas pressure p:

$$\alpha_1 p_1 = \alpha_2 p_2 \tag{3.33}$$

Since V for similar discharges is equal, $E_2 = E_1/a$, and if the pressure $p_2 = p_1/a$:

$$\frac{\alpha_2}{p_2} = \frac{\alpha_1/a}{p_1/a} = \frac{\alpha_1}{p_1} \tag{3.34}$$

so α/p and E/p are invariant. E/p is sometimes referred to as the *reduced pressure* (see Section 2.2). The similarity principle permits comparison of low-pressure discharges under different conditions. Many other similarity relations can be deduced, some of which are listed in Table 3.1.

The term E/p is a measure of the energy gained from the electric field with particle density and hence the average energy gained between collisions. If E/p is constant, similarity conditions hold at low values of electric field and pressure. We can show that $E/p \ll \sigma/e$ [1] if the electric field is low, that is, if the average

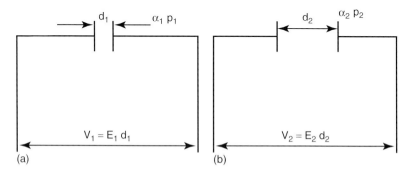

Figure 3.10 Conditions for similarity (a) electrode separation d_1, (b) electrode separation d_2.

Table 3.1 Some similarity parameters.

Townsend coefficient	$\alpha_1 = \alpha_2$
Pressure	$p_1 = \alpha p_2$
Axial dimension	$l = al$
Mean free path	$\lambda_1 = a\lambda_2$
Voltage gradient	$E_2 = aE_1$
Breakdown voltage	$D_1 P_1 = d_2 P_2$
Current density	$J_2 = a^2 J_1$
Electron density	$Ne_2 = a^2 Ne_1$

energy from the electric field is much less than the average energy due to a thermal collision. If E is high, different collision processes occur and the interchange of energy is complex.

References

1. Jewett, J. and Serway, R.A. (2008) *Physics for Scientists and Engineers with Modern Physics*, 7th edn, Thomson Higher Education, Belmont, CA.
2. Fridman, A. (2008) *Plasma Chemistry*, Cambridge University Press, Cambridge.
3. Grill, A. (1994) *Cold Plasma in Materials Application*, IEEE Press, Piscataway, NJ.
4. Cambel, A.B. (1963) *Plasma Physics and Magneto Fluid Mechanics*, McGraw-Hill, New York.
5. Roth, J.R. (1995) *Industrial Plasma Engineering, Vol. 1, Principles*, Institute of Physics, Bristol.
6. Francis, G. (1960) *Ionisation Phenomena in Gases*, Butterworth, London.
7. von Engel, A. (1983) *Electric Plasmas and their Uses*, Taylor & Francis, London.
8. von Engel, A. (1965) *Ionised Gases*, Oxford University Press, Oxford (reprinted 1994, American Institute of Physics, New York).

Further Reading

Brown, S.C. (1959) *Basic Data of Plasma Physics: the Fundamental Data on Electric Discharges in Gases*, MIT Press Cambridge, MA (reprinted 1997, Classics in Vacuum Science and Technology, Springer, New York).

4
Coupling Processes

4.1
Introduction

Coupling energy to a plasma is normally achieved by transferring energy from an electric field to electrons, which then transfer energy by elastic and inelastic collisions and collisionless processes to other atoms and ions. The magnetic field exists corresponding to the flow of charge and external magnetic fields may also be used to aid the energy transfer. The enormous strides made in the development of high-power, high-frequency semiconductors over the last few decades has made the use of high-frequency power sources for plasmas feasible and opened up many new applications of plasmas.

A number of ways of coupling energy to a plasma are illustrated schematically in Figure 4.1a: electrodes in contact with the plasma (direct coupled) or electrodeless [indirectly coupled by induction (b), (c), (d), capacitive (e) or using microwaves (f)].

The plasma power and power density are determined by the application. The coupling process cannot be considered independently of the plasma and factors such as reactor design, volume of plasma and geometry affect it. The coupling requirements can usually be resolved in terms of power and power density for the process.

4.2
Direct Coupling

Direct coupling of electric discharges was the earliest method used and is still the most widely used method today. The process is dominated by the electrodes in addition to the other constraints common to all plasmas. Electrodes are widely used, of necessity with DC, but also with AC at power frequency and high frequencies. Direct coupling is inherently simple, flexible, relatively inexpensive and capable of use over virtually the entire range of operation of plasmas, but is limited by the geometry of the electrodes and contamination from the electrodes. Some of the seminal publications of the 1950s still provide a useful introduction to the topic

Introduction to Plasma Technology: Science, Engineering and Applications. John Harry
Copyright © 2010 WILEY-VCH Verlag GmbH & Co. KGaA, Weinheim
ISBN: 978-3-527-32763-8

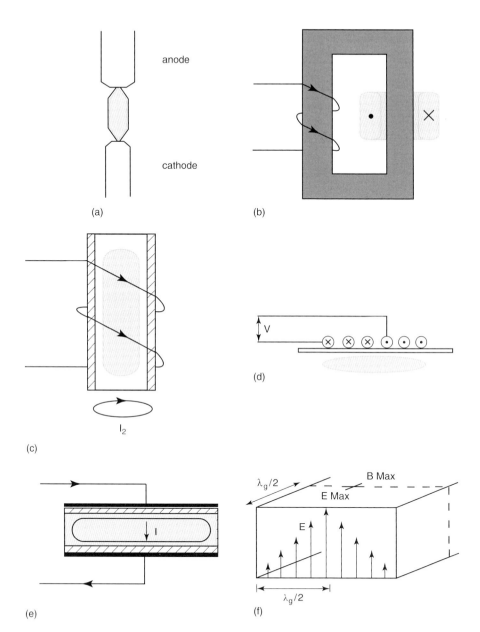

Figure 4.1 Methods of coupling electrical energy to a plasma. Examples of RF inductive and capacitive coupling used in electrodeless reactors. (a) Direct coupling; (b) induction coupling using a ferrous core; (c) induction coil; (d) flat spiral induction coil; (e) capacitive coupling; (f) waveguide.

and there are an enormous amount of empirical results of measurements of the properties of glow and arc discharges, some of which were obtained over 50 years ago and which are still of use today [1–7].

The electric discharge $V–I$ characteristic (Figure 4.2) is widely used to describe the principal regions of the discharge in terms of the measurable variables, voltage and current, and also relates to the appearance of the discharge, but does not include the effects of current density, gas pressure or frequency. (The glow discharge values of voltage and current are for low pressure.)

The discharge characteristic give a clear representation of how the plasma behaviour changes over a wide range of current, but although widely used to describe electric discharges and plasmas, it does not conform with the convention of the independent variable (V in this case) on the x-axis, and the dependent variable I on the y-axis. A more rigorous representation occasionally used is in terms of the variation of J with E.

When a voltage is applied between two electrodes, free electrons between the electrodes from photoionization or gamma-rays are accelerated in the gap over the region O–A in Figure 4.2. As the voltage is increased, electrons accelerated in the electric field collide with neutral particles and a very small current flows until saturation occurs at A and all the free electrons are collected by the anode. When the Townsend region A–B is reached, charge multiplication occurs. If losses are ignored the current is $I = I_0\, e^{\alpha d}$, where α is the first Townsend coefficient and d the electrode separation. If the loss coefficient corresponding to nonionizing collisions is γ, when $e^{(\alpha-\gamma)d} > 1$ the rate of production of electrons exceeds the losses and the current increases over the subnormal region B–C in the glow discharge. Over this region, the current density at the cathode is low and the discharge is diffuse.

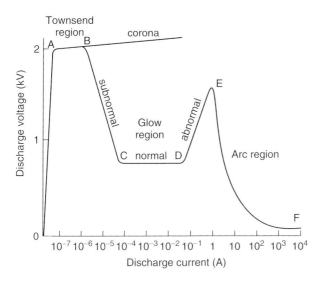

Figure 4.2 The generalized discharge characteristic.

The glow region B–E is a cold nonequilibrium discharge below atmospheric pressure. Over the normal glow region C–D at low pressure, the current is limited by the size of the cathode and in some cases the size of the anode. Very high currents in excess of 500 A are possible if the cathode area is sufficiently large. The abnormal glow D–E corresponds to the limit at which the cathode surface is covered by the glow discharge, and the current density at the cathode increases over the abnormal glow region. As the gas pressure is reduced below about 100 Pa (0.752 Torr), few collisions occur and the glow changes to a beam of charged particles.

A glow to arc transition occurs at E if the area of the discharge at the cathode surface (the cathode root) is insufficient to sustain a glow discharge. The arc discharge E–F is generally in thermal equilibrium and is characterized by high currents and current densities. The cathode voltage drop and the total voltage are generally less than the glow discharge.

The corona discharge at B in Figure 4.2 is not in either charge or thermal equilibrium. Few positive ions are produced in the corona region and, since the electric field strength decreases with radius, the electrons form a unipolar discharge emanating from the active electrode and a high electric field strength region exists close to the electrode and is space-charge limited.

The different parts of the discharge are shown in Figure 4.3. The cathode is the principal source of current, which is mainly carried by electrons. The contribution from the positive ions is small since their velocity is low, but the positive ions help maintain charge equilibrium. The anode acts primarily as a sink for the electrons, although a high voltage gradient usually exists due to the small number of positive ions in the region. The region between the electrodes is known as the *column* and approaches charge equilibrium. The fall regions in front of the electrodes

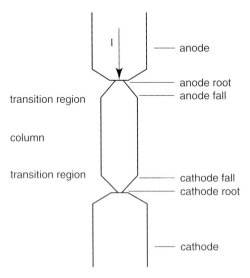

Figure 4.3 The different parts of the electric discharge.

result from the local charge imbalance and their depth, diameter, current density and electric field depend on the mode of the electric discharge. In front of the fall regions are the transition regions corresponding to the transition from the high-field regions in front of the electrodes to the lower field in the column.

The discharge voltage is the sum of the individual voltage drops across the regions: V_c the cathode fall voltage, V_a the anode fall voltage and V_l the voltage drop along the discharge column.

The discharge voltage at constant current can be written as

$$V_c + V_a + V_l = V_d \tag{4.1}$$

The column voltage V_l generally has a negative dynamic resistance which varies with current and can be expressed in the form

$$V_d = V_c + V_a + \frac{C}{I^2} \tag{4.2}$$

where C is a constant.

The cathode fall voltage V_c varies with discharge mode, the work function of the cathode material, gas, current and temperature. The anode fall voltage V_a also varies with the discharge mode, material and gas. The column voltage gradient E_l varies widely with current, gas and gas pressure and impurities.

4.2.1
The Cathode

A plasma or electric discharge is self-organizing, that is, it regulates itself to fit the external parameters. The dimensions of the plasma are controlled by a balance between constriction and expanding forces which vary from being very small in a plasma at low gas pressure to being very high, for example in an arc at high current and atmospheric pressure. The self-regulating effect can be illustrated by a low-pressure glow discharge. The electric field is not in charge equilibrium in the fall region since close to the cathode electrons have insufficient velocities to create positive ions by collision.

The current density at the electrode surface of the cathode is governed primarily by the supply of electrons and determines the discharge mode. Figure 4.4 illustrates the different cathode modes of the glow and arc.

In an electrical conductor such as a metal, diffusion of charges ensures that the conductor is approximately in charge equilibrium. In a plasma the diffusion velocity is lower than in a metal and the plasma takes up a shape which is a balance between magnetic attraction and hydrostatic pressure. The forces affecting the diameter of the discharge are small at low currents and a balance exists between repulsive forces between like charges, for example in the cathode region (where charge imbalance exists), and constriction due to the magnetic field and the cathode fall region in front of the cathode adjusts so as to maintain the flow of current. Over the normal glow region, the size of the cathode root is controlled largely by repulsion of electrons in the cathode fall region where the charge is not in equilibrium, so that as the current increases the cathode root spreads out over the cathode. The

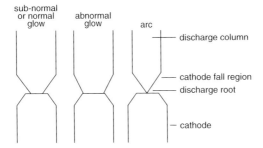

Figure 4.4 Cathode of a glow and arc discharge.

interaction of the self-magnetic field of the plasma with its own current results in an inward radial force (Figure 4.5). The magnetic constricting force per unit length is $F = -\int J_z^2 \mu_0 / 2\pi r$, which is opposed by repulsion between electrons $e^2/4\pi r^2 \varepsilon_0$ [8].

At low currents there will be an optimum diameter of the discharge where the energy losses due to nonionizing collisions is a minimum, and energy from collisions which do not cause ionization assist in the ionization process. (If the diameter is increased, fewer ionizing collisions and more elastic collisions occur; if the diameter of the discharge is reduced, repulsive forces in the electrode regions oppose the change and increase losses in the column.)

In the cathode fall region (and elsewhere in plasmas such as electron or ion beams), charge imbalance occurs and the discharge will expand under the electrostatic forces until the repulsive forces are negligible; however, the force is relatively small and in electron and ion beams at low pressure the beam is easily focused or constricted. Where there is considerable imbalance of charge, for example at an electrode of an arc or glow, the repulsion force may be fairly large but the small radius results in a high magnetic constriction.

In the glow discharge regime, the cathode current density remains constant until the cathode area is covered. When the cathode surface is covered, the current and electric field in the cathode fall region increase to maintain the higher current density and the self-magnetic field increases. When the current density exceeds the

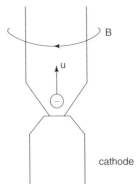

Figure 4.5 Magnetic constriction and divergence of an electric discharge due to the self-magnetic field of the current.

current sustainable by secondary emission the discharge constricts to a value where it can be maintained in the arc mode and is balanced by the increased repulsive forces; the electric field strength increases but the cathode fall voltage decreases.

The self-magnetic field increases with the current density in the electric discharge and interacts with the current to constrict the discharge (Figure 4.5). If the magnetic constriction is greater than the repulsive forces, the discharge constricts to form an arc, increasing the repulsive forces until the forces are again balanced. At the same time, the electric field gradient increases to provide the current density.

The magnetic field associated with the radial component of current also results in an axial force on charges along the axis of the discharge away from the cathode and an axial pressure gradient decreasing away from the electrodes. At high current densities, the pressure gradient causes a jet of plasma and entrained gas from the electrode region along the axis of the discharge. In the column of the discharge where approximate charge equilibrium exists, the constricting force is balanced by the increase in pressure in the column. In the glow discharge, any increase in the radial force may result in a sudden glow to arc transition. Glow and arc modes can exist simultaneously, for example in the low-pressure mercury discharge tube (fluorescent lamp) with a thermionic arc cathode, while the column is in the glow mode.

Magnetic attraction exists between two conductors carrying current in the same direction and the columns of two discharges in close proximity will attract or repel depending on the direction of their current or an external magnetic field.

4.2.1.1 Emission Processes

A conducting surface in a plasma forms a sheath in front of it (see Section 1.3.5). If the conducting surface is connected to an external circuit, it acts as an electrode; the sheath remains but its charge distribution changes. The sheath has fewer electrons at the cathode or positive ions at the anode, which results in high electric field strengths above the electrode surface and particles are accelerated to high velocities in this region. The very small number of electrons in the sheath at the cathode results in fewer excitation and relaxation collisions, causing the characteristic dark space at the sheath interface in front of the surface.

For electrons to be emitted from a material, the energy gap between adjacent electron levels of outer orbits must be close enough to enable electrons to have sufficient energy to overcome the barrier energy ξ_b at the surface of the cathode and break free. The barrier energy is equal to the work function ϕ_w of the cathode material ($\phi_w \approx 4\,eV$ for most metals). A thin adsorbed layer of an electropositive material such as barium or strontium oxide reduces the energy needed to cross the barrier by attracting an electron in the cathode to it. Other forces at the surface vary with the material, the surface condition, adsorbed gases, impurities and contamination and the location of the surface itself is indeterminate due to surface roughness at an atomic level of about 10^{-6} m.

The current density at the surface is related to the Maxwell–Boltzmann energy distribution by the Richardson-Dushman equation

$$J = A_0 T^2 \exp\left(\frac{-e\phi}{kT}\right) \tag{4.3}$$

where A_0 is a universal constant $= 120 \times 10^4$ A m^{-2} K^{-2}, although it can vary over a fairly wide range due to the conditions affecting the local electric field at the surface. The term ϕ is the work function (eV).

Few positive ions are produced in the cathode fall region, which is of the order of the mean free path of an electron λ_e, creating a negative electric field in the dark space in front of the cathode, which is relatively collision free. A high positive electric field is created beyond the dark space towards the cathode by slow-moving positive ions. The positive ions bombard the cathode, releasing electrons which are accelerated to high velocities. At these high values of electric field ($E \approx V/\lambda_e$), the behaviour in the cathode fall is governed by field rather than thermal processes. Many of the electrons travelling through the dark space pass through the space-charge region and recombine with positive ions, resulting in a highly luminous spot in front of the dark space.

Secondary Emission The cathode current density of a glow discharge can be maintained by secondary emission. Secondary emission occurs when electrons are produced by bombardment of a material by electrons, ions or other particles. The particles transfer some of their kinetic energy to neutral particles at the surface which emit electrons. The ratio of the average number of secondary charged particles to the number of equivalent primary charges in collision with the surface is the secondary emission coefficient, δ. The secondary emission coefficients of a very wide range of materials have been measured [4].

Table 4.1 lists secondary emission coefficients for a number of metals commonly used as electrodes or deposition targets for vacuum deposition processes. A high secondary emission coefficient is desirable for a cathode or target for sputtering

Table 4.1 Maximum secondary emission coefficient for various materials.

Metal	δ_{max}	E (eV)
Ag	1.5	800
Al	1	300
Au	1.5	750
Cd	1.1	400
Cu	1.3	600
Fe	1.3	350
Mo	1.25	375
Ni	1.3	550
Pt	1.8	800
W	1.4	700

From Ref. [4].

Table 4.2 Secondary electron emission coefficients.

Substance[a]	$\delta = 1$ at		δ_{max}	V (V) for δ_{max}
	V_1 (V)	V_2 (V)		
Li	–	–	0.5	85
K	–	–	0.7	200
Cu	>100	–	1.3	600
Ag, Au	–	–	1.5	800
W	–	–	1.5	500
C	160	~1000	1.3	600
Soot	–	–	0.4–0.8	500
Pt	–	–	1.6	800
Mo	140	1200	1.3	350
NaCl	~20	1400	6–7	600
MgO	–	–	2.4–4	400
Pyrex glass	30–50	2400	2.3	300–400
Soda glass	30–50	900	~3	300
Oxide cathode BaOSrO	40–60	3500	5–12	1400
ZnS	–	6000–9000	–	–
Ca tungstate	–	3000–5000	–	–

[a] For Na and Zn the four corresponding values are: –, –, 0.8, 300 and 100, 400, 1.1, 200, respectively. From Ref. [9].

whereas a low secondary coefficient is required for an anode or the walls of a reactor. The secondary emission coefficients of a number of other materials used in plasma processes and also electrodes and targets are listed in Table 4.2.

High emission surfaces have low work functions, and the secondary emission coefficient depends on the nature of the surface, impurities, adsorbed gases and so on, which may increase the emission coefficient. The energy of the secondary electrons typically lies in the range 10–20 eV. Surfaces with very thin coatings of materials with high secondary emission coefficients behave in a similar way to coatings used for thermionic emitters. Electropositive coatings result in attraction of electrons in the material closer to the surface and hence a reduction in the barrier energy and may have much higher coefficients.

The slow-moving positive ions create a space charge in front of the cathode and an electric field given by the Langmuir–Child equation [6]:

$$J = \frac{4}{9}\left(\frac{2e}{m_i}\right)^{\frac{1}{2}} \frac{\varepsilon_0 V^{\frac{3}{2}}}{x^2} \qquad (4.4)$$

where d is of the same order as the depth of the cathode fall. An increase in the number of electrons emitted reduces the space charge. A similar space charge, although normally smaller, is formed at the anode also due to the deficiency of positive ions in this region. Solving for V gives an electric field strength of about

10^4 V mm^{-1} at the cathode. The cathode fall voltage $V \propto I^{2/3}$ over the space-charge limited region and the fall region therefore has a positive characteristic.

At low pressures, removal of material may occur by impact of high-energy ions, referred to as *sputtering*. Sputtering is a useful method of coating at low pressures using an electric discharge or plasma, but where unwanted sputtering by high-energy particles may occur, the sputtering yield should be low. The sputtering threshold is within the range 10–50 eV.

Field Emission The Schottky effect [10] is due to the electric field that occurs below the cathode surface. The region in the cathode very close to the cathode surface has different properties from the bulk material in a similar way to a junction of a semiconductor. If a charge is a distance x above the cathode surface, a mirror charge of opposite polarity and equal distance below the plane exists (see Chapter 9).

The force between the charges is

$$F = \frac{e^2}{4\pi \varepsilon_0 (2x_0)^2} \tag{4.5}$$

and the potential energy by integrating from x_0 to infinity

$$\int_{x_0}^{\infty} F dx = \frac{e^2}{16\pi \varepsilon_0 x_0} \tag{4.6}$$

The potential to reduce the force to zero at x_0 is $\frac{e^2}{16\pi \varepsilon_0 x_0} - e\phi = 0$

Giving

$$x_0 = \frac{e}{16\pi \varepsilon_0 \phi} \tag{4.7}$$

The voltage between the surface and the charge may be only a fraction of a volt, but this is sufficient to reduce the effective work function and if the electron has sufficient energy to reach the cathode surface it will be accelerated in the cathode fall region.

If the Schottky effect is taken to extremes, the barrier becomes so thin that it is possible for electrons to tunnel through it at electric fields of the order of 10^8 V m^{-1}. This can occur at moderate voltages around wires and even plane electrodes and is one of the reasons combined with local concentration of electric stress at the surface for the filamentary discharges observed between some barrier electrodes with active negative electrodes.

Thermionic Emission The energy gap or work function ϕ (between adjacent electron levels of outer orbits) is close enough for electrons to be released from a solid cathode by thermionic emission at temperatures greater than $\sim$2000 °C. The effect of heating the cathode is to reduce the energy required to free an electron in a way analogous to evaporation. Electrons may be emitted from a solid but at room temperatures the number emitted is inconceivably low (one electron in about 10^{14} years), but at higher temperatures the emitted current density may be over 10^3 A m^{-2}. Some representative thermionic emission data is given in Table 4.3.

Table 4.3 Representative thermionic emission data.

Metal	$A_0 (m^{-2} K^{-2})$	Work Function (eV)
W	70.0×10^4	4.55
Ta	55.0×10^4	4.25
Ni	30.0×10^4	5.15
Cs	160.0×10^4	2.14
Pt	32.0×10^4	5.65
Cr	48.0×10^4	4.50
Ba on W	1.5×10^4	1.56
Cs on W	3.2×10^4	1.36
C	–	5.0

From Ref. [8].

Metal oxides generally have higher melting points than metals and some have lower work functions. The work functions of some metal oxides are listed in Table 4.4.

The Stefan–Boltzmann equation [Eq. (3.34)] indicates that only the temperature and the work function have an affect on thermionic emission. Halving the value of the work function ϕ from 4 to 2 increases the current density over seven times at 2000 K; an increase in temperature of 50% from 2000 to 3000 K increases the current density by about two and a half times the value at 2000 K and $\phi = 4$. The implications of this are that in practice thermionic emission is governed mainly by the temperature of the cathode.

Table 4.5 lists the current densities of different emission processes as a function of electric field strength. The most common cathode materials are aluminium at low pressures for glow discharges, arcs use thoriated tungsten for thermionic emission and copper for cold cathodes, and anodes and nickel for anodes in low-pressure glow discharges. Strontium barium titanate is used for hot cathodes in fluorescent

Table 4.4 Work functions of metal oxides at high temperatures.

Oxide	Work function (eV)
BaO	1.0
CuO	5.2–5.4
Cu_2O	5.2–5.4
Th_2O_2	6
WO_3	3.5
ZrO_2	9.2

After Ref. [11].

Table 4.5 Current densities of thermionic, field and thermionic field emission at different electric fields E.

Electric field × 10^6 (V cm^{-1})	Schottky decrease of W (V)	Thermionic emission j (A cm^{-2})	Field emission j (A cm^{-2})	Thermionic field emission j (A cm^{-2})
0	0	0.13×10^3	0	0
0.8	1.07	8.2×10^3	2×10^{-20}	1.2×10^4
1.7	1.56	5.2×10^4	2.2×10^{-4}	1.0×10^5
2.3	1.81	1.4×10^5	1.3	2.1×10^5
2.8	2.01	3.0×10^5	130	8×10^5
3.3	2.18	6.0×10^5	4.7×10^3	2.1×10^6

Electrode temperature, work function, Fermi energy and pre-exponential Sommerfield factor are $T = 3000$ K, $W = 4$ eV, $\varepsilon_F = 7$ eV and $A_0(1 - R) = 80$ A cm^{-2} K^{-2}, respectively. From Ref. [12].

lamps. A summary of the different cathode emission processes is included in Table 4.5 as a function of the electric field strength at the cathode surface.

4.2.2
The Cathode Fall Region

The cathode fall voltage of a cold cathode glow discharge over the normal glow region is due to the space charge in front of the cathode and is a function of the electrode material, gas and gas pressure and varies widely. Table 4.6 lists some values of the depth of the cathode fall at low pressures for different materials and gases in terms of the similarity parameter dp (see Section 3.4). Typical values of

Table 4.6 Normal cathode fall thicknesses in cm Torr at room temperature.

Cathode	Air	Ar	H$_2$	He	Hg	N$_2$	Ne	O$_2$
Al	0.25	0.29	0.72	1.32	0.33	0.31	0.64	0.24
C	—	—	0.9	—	0.69	—	—	—
Cu	0.23	—	0.8	—	0.6	—	—	—
Fe	0.52	0.33	0.9	1.30	0.34	0.42	0.72	0.31
Mg	—	—	0.61	1.45	—	0.35	—	0.25
Hg	—	—	0.9	—	—	—	—	—
Ni	—	—	0.9	—	0.4	—	—	—
Pb	—	—	0.84	—	—	—	—	—
Pt	—	—	1.0	—	—	—	—	—
Zn	—	—	0.8	—	—	—	—	—

From Ref. [13].

Table 4.7 Comparison of arc roots on thermionic and cold cathodes.

Parameter	Thermionic	Non-thermionic (cold cathode)
Cathode fall region		
Voltage (V)	10	15
Depth (mm)	0.001	0.1
Electric field (V mm^{-1})	10×10^3	150
Cathode root		
Temperature (K)	2800	2300
Movement	Stationary or slow movement	Rapid movement
Current density (A mm^{-2})	$1-10^2$	$10^4 - <10^6$

cathode fall voltages vary over several hundred volts depending on the cathode material, geometry, gas, impurities and gas pressure [13].

The cathode fall voltage for an arc is different for thermionic and cold cathodes. The cathode fall voltage of an arc at atmospheric pressure is about 10 V for thermionic emitters and up to about 15 V for cold cathode materials at atmospheric pressure. The electric field strength in the cathode fall region of a cold cathode is of the order of 4×10^6 V mm^{-1}; the corresponding current densities are of the order of 0.1–1 A mm^{-2} for a thermionic emitter and up to 10^4 A mm^{-2} for a cold cathode. The depth of the cathode fall of a cold cathode is less than 0.1 mm. Table 4.7 summarizes the principal differences between thermionic and cold cathode emission processes.

4.2.3
The Anode

The principal function of the anode of an arc is to act as a sink for the electrons arriving from the column. Although some positive ions are produced by impact from the electrons and also sputtered atoms, a negative space-charge sheath exists due to the small number of positive ions in the anode fall. Table 4.8 gives some values for the anode fall voltage and anode fall depth at low pressures. The voltage drop over the anode fall and the power density is generally lower than that at the cathode but the total power transfer is often larger due to the higher total heat flux along the axis of the discharge which is less constricted at the anode.

4.2.4
The Discharge Column

The column of the discharge satisfies the conditions for a plasma, that is, it is in charge equilibrium over a region greater than the Debye radius and the collision frequency is much larger than the plasma frequency. The dimensions of the

Table 4.8 Anode fall voltage, V_A, and anode fall thickness, d_A.

Gas	Pressure (Torr)	V_a (V)	d_a (cm)
H$_2$	1–5	17–19	0.5–0.7
N$_2$	0.8	16.5	–
	0.5	15.7	–
O$_2$	1	14.2	–
	~0.1	–	>0.12
He	10	26	–
Ne	10	17	–
Ar	20	10	–
	0.5	15.3	–

After Ref. [13].

discharge column are defined by the length of the discharge path (not necessarily the distance between electrodes) and the balance of repulsive and constricting forces.

The column area of cross-section in a glow discharge is dependent only on the current unless it is enclosed, for example in lamps or lasers, where it is stabilized by the walls, and typically will be in the abnormal region of the glow discharge so as to have the highest current and current density. The current density in the column varies with gas pressure, gas and additives and the colour varies with the gas.

At low gas pressures, the arc column may operate in a diffuse mode since the repulsive forces are significant but at higher pressures close to atmospheric the arc constricts.

Figure 4.6 illustrates the variation of electron temperature and gas temperature with gas pressure for an arc up to atmospheric pressure. At low pressure the electrons may have much greater energies than the ions and the magnetic

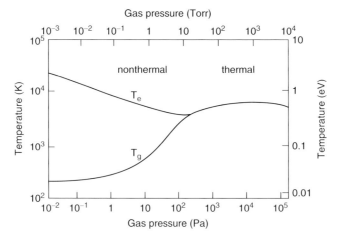

Figure 4.6 Variation of electron and gas temperature with pressure. (From Ref. [13].)

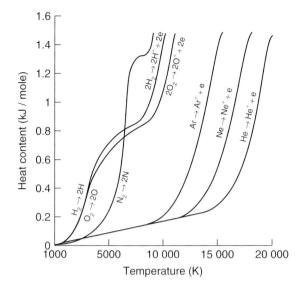

Figure 4.7 Variation of heat content of gases with temperature. (From Ref. [5].)

constricting force is opposed by repulsion of like charges. The current density in the arc column is governed by the transport properties of the gas, the current density and cooling. The current density in the arc column is normally much higher than in a glow discharge and varies with gas pressure. In the arc column at atmospheric pressure, a balance exists between gas pressure and electromagnetic constricting forces.

The variation of the heat content with temperature of a number of atmospheric gases at atmospheric pressure is shown in Figure 4.7. The graphs show the effects of dissociation and ionization. The current density in the arc column is governed by the transport properties of the gas and the current density and cooling, and peaks in the thermal conductivity of gases account for phenomena such as high-temperature arc cores.

The arc is normally mainly cooled by convection, which can be increased by forced convection. The effect of convection decreases at high altitude due to the reduced force of gravity and can be demonstrated by spinning the discharge, which imposes a centrifugal force on the ions that expands the discharge column (Figure 4.8).

The voltage gradient and hence temperature and power (assuming a constant current supply) can be increased, for example by cooling by forced convection at atmospheric pressure, such as in the plasma torch, or by constriction in compact fluorescent lamps.

4.2.5
Interaction of Magnetic Fields with a Discharge or Plasma

The interaction of a magnetic field with a discharge or plasma in charge equilibrium is the same for an electron or ion since although the charges are of opposite polarity,

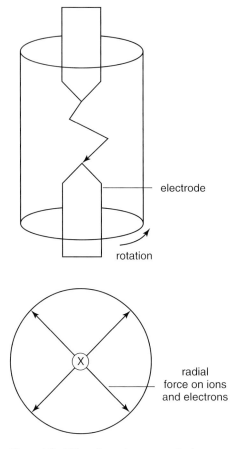

Figure 4.8 Effect of rotating an arc discharge.

the direction of velocity is the same. The magnetic force is opposed by aerodynamic forces and in addition a back EMF is induced in the plasma, $e = -d\phi/dt$.

Figure 4.9 shows different configurations of discharges in magnetic fields. The magnetic driving force from the interaction of the self-magnetic field of the current in the conductor and the discharge current which are at right-angles to each other is shown in (a). This force is frequently used in switchgear but is a nuisance in welding. The force can be nullified (b) by balancing the current flow. A driving force can also be produced by an external magnetic field transverse to the discharge (c) or an axial magnetic field perpendicular to a discharge between coaxial electrodes (d) to rotate the discharge. If both the discharge and magnetic field share the same axis, any off-axis component experiences a radial force and a rotational velocity used in vacuum arc switchgear, the vacuum arc furnace and magnetron.

A summary of the distinguishing characteristics of low-pressure glow and atmospheric arc discharges is given in Table 4.9.

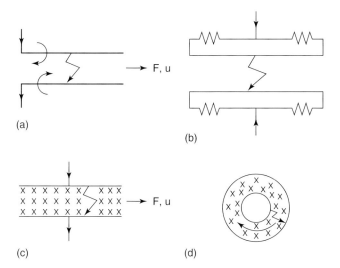

Figure 4.9 Effect of a magnetic field of an arc.
(a) Magnetic forces acting on a discharge due to its own self-magnetic field; (b) balanced current flow; (c) in a transverse magnetic field; (d) rotation between coaxial electrodes in a transverse magnetic field.

Table 4.9 Distinguishing characteristics of low-pressure glow and atmospheric pressure arc discharges.

Discharge region	Representative values	
	Glow	Arc
Cathode process	Secondary emission	Thermionic/field emission
Fall voltage (V)	300	8–15
Fall thickness (mm)	10	0.1–10
urrent density (A mm^{-2})	0.1	10–10^4
Anode		
Fall voltage (V)	20–30	3–12
Fall thickness (mm)	10	0.1
Current density (A mm^{-2})	0.1	10
Column		
Voltage gradient (V mm^{-1})	1–10	1
Current density (A mm^{-2})	2×10^{-3}	10
Mean temperature of neutral particles (K)	Close to ambient	6000
Number density (electrons m^{-3})	5×10^{15}	–

4.3
Indirect Coupling

Indirectly coupled discharges enable a plasma to be obtained without the constraints of electrodes. The availability of high-frequency power sources using electronic valves led to increased interest and applications such as spectroscopy, and more recently semiconductor supplies have led to the widespread application of electrodeless plasmas in areas such as the manufacture of semiconductors. The use of magnetic fields to control the behaviour of the plasma specifically by delaying its transit through the energizing electric field by spinning it around magnetic field lines has enabled high-energy densities to be obtained and charged particles can be directed at a target to be obtained even at low pressures [14].

Coupling in the near field is analogous to direct coupling using electrodes where the pathlength corresponds to the that of the column of a directly coupled discharge but without the electrode regions. The *near field* is loosely defined as where the dimensions of the electrodes and the length of the conducting path is not more than $\lambda/10$ of the wavelength of the supply frequency. Over the near field, the magnetic field causes induced voltages in nearby conductors and is responsible for the skin effect. Over this region, the effect of propagation can be largely ignored and the components and connectors considered in terms of discrete values rather than distributed values and the effect of the far field subsumed in the near field.

4.3.1
Induction Coupling

In an inductively coupled plasma (ICP) (H field), energy is coupled by inducing a current in a similar way to the secondary winding of a transformer (Figure 4.10a) with the plasma acting as a single turn.

The current and voltage in the plasma loop can be determined from the transformer equations

$$\frac{V_1}{V_2} = \frac{N_1}{N_2} \text{ and } I_1 N_1 = I_2 N_2 \text{ where } N_2 = 1 \qquad (4.8)$$

The power coupled in the secondary is a function of the magnetic flux density and frequency, so that at low frequencies a ferromagnetic core is used to increase the coupling. Up to about 20 kHz an iron core is used; ferrite cores can be used up to more than 400 kHz. Higher frequencies can be coupled with an air core. Figure 4.10a shows a transformer with a single-turn secondary used to couple energy to a toroidal plasma. The transformer also serves to match the single-turn plasma to the supply. The power dissipated in the plasma and the uniformity of the distribution are affected by the skin depth and the geometry. The diameter of the cross-section of the plasma formed by a toroidal plasma has an optimum for maximum power and energy density between low power densities at small cross-section and high power at low power densities at large cross-section. A compromise at about twice the skin depth is sometimes appropriate. In the case of the toroid, the radial depth of the conducting region which is in the plane of the enclosed area is about 2δ.

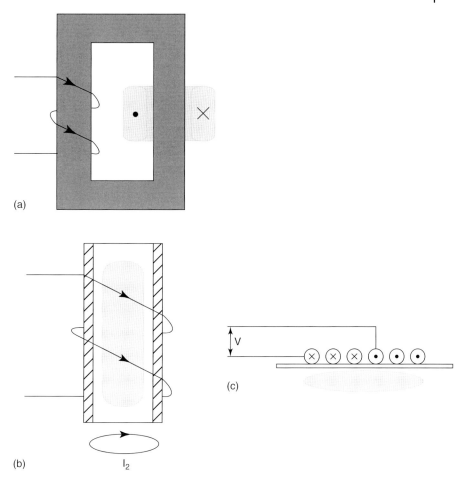

Figure 4.10 Examples of induction coupled plasmas:
(a) transformer; (b) solenoidal coil; (c) flat spiral.

The effect of high-frequency electromagnetic fields was extensively studied by Babat in 1947 [6] and graphically demonstrated by Reed in 1961 [7]. The solenoid coil (Figure 4.10b) produces an approximately annular plasma [7, 15]. The depth of the annular plasma is affected by the flux from the circulating current component diametrically opposite which is in the opposite direction. The effect is small for $\delta \leqslant D/10$, where D is the outer diameter of the annular plasma.

If the collision frequency ν_c is much greater than the supply frequency ω_s, particles do not have time to respond and power absorption occurs in the skin depth in a similar way to a metal. If the collision frequency is much less than the supply frequency, the wave may be transmitted like light through a transparent material.

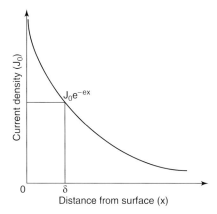

Figure 4.11 Attenuation of current density with distance from the surface due to the skin effect.

The amplitude of an incident wave in an infinite medium decays to 1/e of the value at the surface at the skin depth [14] (Figure 4.11):

$$\delta = \left(\frac{2}{\omega \sigma_s \mu_0}\right)^{\frac{1}{2}} \quad (4.9)$$

by writing $\sigma = e n_e (u_d/E)$ (Eq. 2.3) and $u_d = (eE/m_e)\tau_{ce}$ (Eq. 2.6) [14]:

$$\delta = \sqrt{\frac{2m_e}{e^2 \mu_0}} \sqrt{\frac{\nu_{ce}}{n_e \omega}} \quad (4.10)$$

The first term in (4.10) is constant for a given gas so the skin depth $\delta \propto \sqrt{\frac{\nu_{ce}}{n_e \omega}}$. For a finite plasma, the skin depth is an accurate measure of penetration of the current for

$$\frac{D}{\delta} \geq 10 \quad (4.11)$$

For maximum power dissipation in the plasma, the width of the plasma channel is about twice the skin depth.

The flat spiral coil in Figure 4.10c produces a uniformly distributed plasma over a large area.

4.3.2
Capacitive Coupling

A capacitively coupled plasma is represented by a lossy dielectric and is similar after breakdown to a discharge in series with capacitors corresponding to the insulated layers (Figure 4.12) [12].

Capacitive coupled plasmas behave in a similar way to direct coupled discharges at high frequencies. Energy is coupled to the plasma between electrodes covered by a layer of insulation with the plasma in the space between them. The insulation

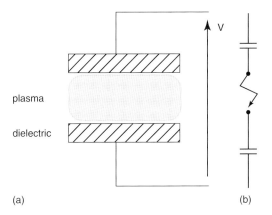

Figure 4.12 Capacitively coupled plasma: (a) schematic and (b) equivalent circuit after breakdown.

acts as a distributed stabilizing element, reducing the tendency of the discharge to constrict and protecting the electrodes from sputtering or evaporation.

The current flowing in a capacitor is due to the alternating charge polarity at the surfaces of the plates of the capacitor, and current flows across the gap due to the displacement of the charge on the dielectric each half cycle. A component of current flows in addition to the displacement current, due to lossless dipolar and rotational excitation in the gap. Electrons are accelerated and oscillate in the electric field and if the field is high enough ionization occurs.

The low values of inductance and capacitance required at high resonant frequencies (>30 MHz) necessitate the use of resonant cavities and waveguides rather than discrete circuit components due to the inherent values and parasitic effects of the components and their connections. A high-frequency resonant circuit can be produced by a box-shaped cavity. The two opposite sides of the cavity form the capacitor and the other two sides act as the inductor.

4.3.3
Propagation of an Electromagnetic Wave

Propagation of a wave in a plasma enables the limitations of coupling electromagnetic energy by electrical induction in the near field, which is effectively limited by the skin effect, to be overcome particularly at low particle densities. The equation for the skin depth (Eq. 4.10) is a good measure of whether energy is being effectively absorbed in a plasma and shows that the only variables are the supply frequency, the collision frequency and the electron number density. To obtain a high number density of electrons, a high collision frequency and low supply frequency are necessary, which are incompatible at low pressures.

At low pressures, it becomes increasingly difficult to achieve a high enough number of electrons of sufficient energy to in turn transfer their energy to heavy particles so as to carry out processes such as deposition on a practical scale. High frequencies permit coupling in the particle regime where fewer collisions occur, but higher particle energies and higher energy densities can be obtained.

Higher electron number densities can be obtained by propagating an electromagnetic wave into a plasma at the plasma resonant frequency. An electromagnetic wave is dispersed at an interface with a medium of different permittivity in the same way as light. Attenuation, transmission and reflection can occur.

Propagation of an electromagnetic wave is described by the dispersion relation [10]

$$\omega^2 = c^2 k^2 + x \tag{4.12}$$

where ω is the frequency of the electromagnetic wave, c_0 is the velocity of light, x is a constant depending on the medium and the wavenumber $k = 2\pi/\lambda$. This is illustrated in Figure 4.13. For a conducting medium such as an unmagnetized plasma:

$$\omega^2 = c^2 k^2 + \omega_p^2 \tag{4.13}$$

and the wave will be transmitted for

$$\omega > \omega_p \tag{4.14}$$

For the condition $\omega \leqslant \omega_p$, the wavenumber is zero and the wave will be reflected. A similar case exists for ions. The assumption that the wave will be reflected is not completely true since first there is interaction between resonance by ions with the electrons and since the boundary of a technological plasma is likely to have a lower electron number density than at the centre the refractive index decreases as the wave penetrates the plasma until it is reduced to zero or reflected. The above indicates, however, that unlike waves in the ionosphere the range of frequency over which attenuation occurs is small in technological plasmas.

The effective permittivity ε of the plasma is

$$\varepsilon = \varepsilon_0 - \frac{n_e e^2}{m\omega^2} \tag{4.15}$$

so that the permittivity of a plasma is less than ε_0 and the wave diverges away from the normal. The electron density n_e increases from zero at the boundary,

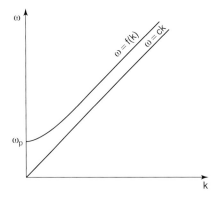

Figure 4.13 Variation of the frequency ω of an electromagnetic with wavenumber k.

whereas the refractive index decreases and the dispersion increases from zero to a maximum at the maximum electron density (Figure 4.14) and will be reflected if $n_e > n_c$. The wave may be reflected if the electron density increases sufficiently or eventually be attenuated to zero.

Energy exchange by Landau damping occurs when an electromagnetic wave is propagated in a plasma. If the electrons in a magnetized plasma are rotating at their gyro frequency at a frequency close to the frequency of the incident wave, resonance occurs. Below the phase velocity of the wave, the particles with gyro velocities less than the phase velocity will be accelerated whereas those with particles slightly above will be decelerated and lose energy. If the plasma resonant frequency is the average velocity and the velocity distribution is Maxwellian, the number of particles below the resonant frequency is greater than that above and there is a net gain in energy.

At the high frequencies used, dimensions of connecting wires and components become comparable with the supply wavelength (2.45 GHz = 0.122 m). At these wavelengths, the signal is propagated by an aerial or waveguide. The propagation of waves in magnetized plasmas is complex and dealt with elsewhere [17] but can be considered in terms of simple models.

Below the resonant frequency in the particle interaction region the plasma is able to screen the wave and the wave is unable to propagate and decays exponentially or is reflected if the electron density is high enough for collective behaviour.

When the plasma frequency is reached, energy is interchanged between the plasma and the plasma resonates.

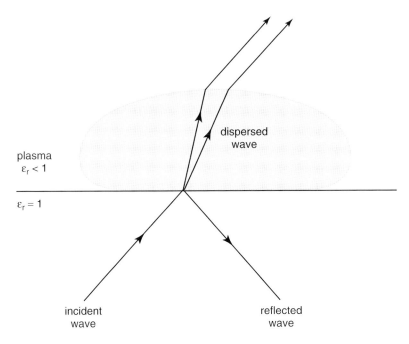

Figure 4.14 Reflected wave at an interface with a plasma.

If the wave frequency is greater than the plasma frequency, the plasma is unable to respond fast enough and the wave is propagated without attenuation. In a weak plasma, however, the plasma frequency $\omega_p \propto \sqrt{n_e}$ and if the electron density increases until collective behaviour occurs when the wave frequency is equal to the plasma frequency the wave is reflected.

4.3.4
The Helical Resonator

One way of increasing the energy absorbed by a plasma at low pressures is to increase the length of time spent during which energy transfer occurs. This is achieved in the helical resonator [12], which, although similar to an ICP, has a slow wave construction so that it achieves longer residence times without the need for the magnetic field produced by the solenoid used in electron cyclotron resonance (ECR).

The helical resonator (Figure 4.15) consists of a helically wound coil. The length of conductor is a quarter or half the wavelength of the resonant frequency with turns spaced apart by the wire width. The coil is surrounded by a grounded screen which forms a transmission line with distributed capacitance and also avoids parasitic capacitors and reflections due to mismatching, and forms a resonant cavity at quarter or half wave resonance. The skin effect is minimized by plating the shield and walls and helix with high conductivity metal and Q factors in excess of 1000 are obtained.

The velocity of the wave moving along the axis of the helical coil is governed by the pitch of the winding. The supply is connected by a matched line to the coil. At the high frequencies used (600–1500 MHz, 0.5–0.2 m) the flux is perpendicular to the axis of the coil and the E field is in an azimuthal direction in the opposite direction to the current in the coil and the coil acts as a dipole resonator to produce a travelling wave, which causes the current to spiral down the helix and oscillate back and forth from one end of the coil to the other at its resonant frequency.

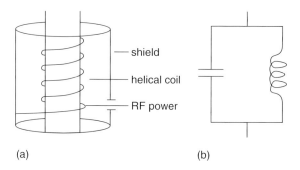

Figure 4.15 Schematic of a helical resonator: (a) schematic and (b) equivalent circuit.

4.3.5
Microwave Waveguides

The electric field in a waveguide (Figure 4.16) can be sufficient to cause electrical breakdown at low gas pressure and form a plasma [18]. A waveguide can be considered in terms of a resonant cavity formed by a hollow conductor of square section. If the two opposite sides are removed to leave two parallel strips, the current flows along the strips, and an E field exists between them, and the magnetic field surrounds the current (see Chapter 9). The distribution of the electromagnetic field in the waveguide is described by the transverse electromagnetic (TE) mode which indicates that the wave is a TE wave and suffixes indicate the mode structure being propagated and correspond to the number of half periods of electric field distribution along perpendicular sides of the waveguide. The TE_{10} electric field has a maximum at the centre of the wide wall perpendicular to the E field. For the ratio $a/b = 2$ the electric and magnetic fields alternately reach maximums at $\lambda_g/4$ apart along the axis of the waveguide.

The wavelength is determined by the geometry which in a rectangular waveguide is normally the width (a) and height (b) (Figure 4.16). The cut-off frequency is given by Roussy and Pearce [19] as

$$\omega^2 \mu \varepsilon \geqslant \left(\frac{m\pi}{a}\right)^2 + \left(\frac{n\pi}{b}\right)^2 \tag{4.16}$$

where m and n are integers. TE_{10} $m = 1$ and $n = 0$ give the smallest size for a given frequency.

Rearranging Eq. (4.16), the cut-off wavelength is

$$\lambda_c = \frac{c}{f_c} = \frac{1}{\sqrt{\mu \varepsilon_0} f_c} = \frac{2}{\sqrt{(m/a)^2 + (n/b)^2}} \tag{4.17}$$

The waveguide is normally at atmospheric pressure and a tube containing the plasma gas at low pressure passes through the region of high electric field or the wave is transmitted through a dielectric window into a reactor. A circular cavity may also be used to couple the waveguide to the plasma (Figure 4.17) by passing a tube through the region in the waveguide where the E field is a maximum. If a

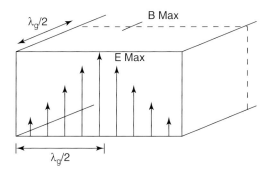

Figure 4.16 Electromagnetic field distribution in a rectangular TE mode.

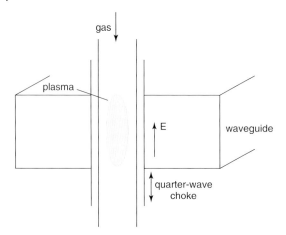

Figure 4.17 Plasma coupled to a waveguide.

resonant cavity is used, the radius r and depth d of the coupled region are made comparable to the wavelength.

The wavelength in the waveguide depends on the effective pathlength between reflections $\lambda_g/2$ but the frequency is constant. The TE_{10} mode is the most widely used but many other possible modes exist. To prevent other modes being propagated, the cut-off frequency is chosen to be just below the operating frequency. Matching the supply and source is important since if it is not matched correctly the reflected wave may damage the source where the wave leaves a waveguide through an aperture the impedance of free space is taken as 377 Ω. Reflections are reduced by a dummy load, matching and the circulator.

4.3.6
Electron Cyclotron Resonance

An electron cyclotron resonator is illustrated in Figure 4.18. The axial magnetic field interacts with electrons in the solenoid which rotate at their gyro velocity [19]. As the plasma electron resonant frequency is approached, the average power absorbed by a volume containing n_e electrons becomes

$$P_{av} = \frac{n_e e^2 E_0^2}{2m} \left[\frac{\nu}{\nu^2 + (\omega_s - \omega_c)^2} \right] \qquad (4.18)$$

The power input increases rapidly to a maximum at the plasma resonant frequency resonance when $\omega_s = \omega_c$ electron cyclotron resonance occurs. Either side of the resonance frequency, the coupling decreases in a similar way to the Q of a resonant electrical circuit.

Figure 3.6 shows the variation of gyro frequency with magnetic flux density. Figure 4.19 shows the variation of the electron number density with magnetic flux density at the electron cyclotron frequency. The variation of the electron number density with the frequency of the power supply at the electron cyclotron frequency

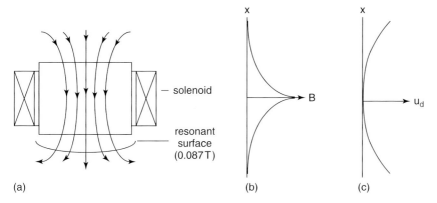

Figure 4.18 Electromagnetic field configuration used in the ECR. (a) magnetic field, (b) variation of magnetic flux density along axis of the solenoid, (c) variation of axial velocity along axis of solenoid.

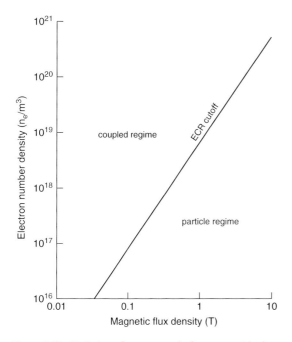

Figure 4.19 Variation of power supply frequency with electron number density.

is shown in Figure 4.20. The electron number density in the plasma increases with frequency but the choice of frequency is limited to the ISM frequencies which on pratical grounds limits it to 2.45 GHz. The magnetic flux density to obtain a gyro frequency of 2.45 GHz is 0.087 T (870 Gauss). At 2.45 GHz and 0.087 T the corresponding electron density is 7.44×10^{16} m^{-3}.

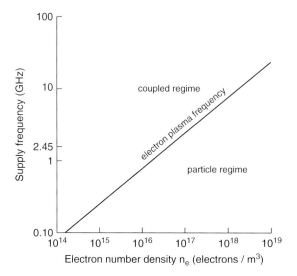

Figure 4.20 Variation of electron frequency with magnetic flux density.

From Eq. (3.18), the electron number density at the electron cyclotron resonant frequency is

$$n_{ce} = \frac{\omega_{pe}^2 m_e \varepsilon_0}{e^2} = \frac{4\pi^2 m_e \varepsilon_0}{e^2} v_{ce}^2 \tag{4.19}$$

If the electron number density n_e is specified and the plasma frequeny ω_{pe} is known the cyclotron frequency ω_{ce} can be adjusted by varying the magnetic flux density so that $\omega_{ec} = \omega_p$. Substituting for constants and solving for $f_{pe} = \omega_{pe}/2\pi$, $n_{ce} = 0.0124 f_{pe}^2$, and the plasma frequency is conveniently given by

$$f_{pe} \approx 9 n_e^{\frac{1}{2}} \tag{4.20}$$

If

$$\omega_e \leqslant \frac{\varepsilon_0 B}{m_e} \tag{4.21}$$

the incident energy will disperse, and if

$$\omega_e \geqslant \frac{\varepsilon_0 B}{m_e} \tag{4.22}$$

it will penetrate the plasma (see Section 4.2.3).

If the gyro frequency ω_{ce} is below the electron resonant frequency ω_{pe}, $\omega_{ce} < \omega_{pe}$, the particles can respond individually to changes in electric field and can experience collisions and absorb energy which is enhanced by resonance (particle regime).

If the cyclotron frequency ω_{ce} is above the electron plasma frequency ω_{pe}, the electrons can no longer respond individually due to their inertia and they behave collectively. The signal is reflected but also penetrates the plasma over a limited distance, is dispersed and is attenuated exponentially. At very high frequencies and low pressures, for example 1 Pa (7.5×10^{-3} Torr), the energy input is enhanced by oscillation of the charged particle at its resonant frequency (plasma frequency) and absorption occurs.

From the Loschmidt number ($n_L = 2.69 \times 10^{25}$ particles m^{-3}) at a pressure of about 1 Pa (7.52×10^{-3} Torr) about 10^{20} neutral particle density and a degree of ionization of 1% the electron number density is about 10^{18}. Figure 4.18 indicates that operation in the particle regime can only be achieved within the microwave band at 2.45 GHz at about 10^{-2} Pa (75.2×10^{-6} Torr). Higher pressures require higher frequencies, which in general is not practical although, for example, lasers can be used to break down air at atmospheric pressure and are used in inertial fusion research. Figure 4.20 indicates by extrapolation that since it is difficult to achieve electron number densities greater than 10% of the particle density, this corresponds to 2.54 GHz at 10^{-4} Pa ($\sim 10^{-6}$ Torr) and at atmospheric pressure a frequency of 10^5 GHz.

The cyclotron enables electron densities as high as 10^{18} m^{-3} to be achieved.

Power densities at very low pressures down to 10^{-2} Pa (7.5×10^{-4} Torr) can be obtained by coupling energy to electrons at the electron resonant frequency. The plasma is produced in a waveguide and fed through a window into a solenoid. Since the free space wavelength at 2.45 GHz is 122 mm, a reactor vessel 0.1 m in diameter is still within the near field.

The electric field is perpendicular to the magnetic field so that the electrons in the plasma rotate around the flux lines at the gyro frequency in the same way as in the magnetron:

$$r = \frac{m_e u}{eB} \tag{4.23}$$

The solenoid is energized so that it is right polarized, hence the radial component of the diverging magnetic field results in the electrons propagating along the diverging magnetic field with decreasing axial velocity (see Section 3.2.1) and increasing gyro velocity into the solenoid. If the axial velocity is sufficient when the centre of the solenoid is passed, the axial velocity increases and the gyro velocity decreases. The ions restrain the axial movement of the electrons and electrons and ions are produced by collision and collisionless transfer.

The direction of the magnetic field determines the direction of rotation of the electrons. When the direction of the particle is such that the current results in an increase in the magnetic field (right polarized) in the same direction as the gyromotion, the charged particles gain energy from the electric field.

As they leave the solenoid, they pass through a surface where the flux density is the optimum for electron resonance at the gyro frequency [0.0875 T (875 G)] and

$$\omega_{ce} = \omega_{pe} \text{ or } \frac{eB}{m_e} = \left(\frac{e^2 n_e}{\varepsilon_0 m_e}\right)^{\frac{1}{2}} \tag{4.24}$$

When the electrons reach the magnetic flux for resonance, energy coupling to the ions from the electrons increases and the effect of the diverging magnetic field on the compressible ions is to cause a longitudinal acoustic wave.

The cut-off frequencies in practice are not as well defined as the equations suggest and some degree of interaction between the ions and electrons occurs so that electron resonance occurs at an upper hybrid resonance frequency:

$$\omega_h = \left(\omega_{ce}^2 + \omega_{pe}^2\right)^{\frac{1}{2}} \qquad (4.25)$$

which is greater than the gyro frequency.

ECR is used in chip fabrication but also in space vehicles using large numbers of ions ejected at high velocity to develop thrust and ion cyclotron resonance is used in Tokamak fusion reactors.

4.3.7
The Helicon Plasma Source

Helicon waves can penetrate a plasma and are not limited to the skin depth like TE waves [20]. Helicon waves are used during the manufacture of semiconductors and in fusion research, for propulsion of vehicles in space and are also encountered in communication. A helicon plasma reactor is capable of producing a high-energy density plasma from an RF source at a lower magnetic field strength than ECR, is relatively compact and enables very high number densities of ions ($n_i > 10^{17}$ m^{-3}) to be obtained at low gas pressures.

The helicon wave is named after its helical path of propagation and is characterized by its permittivity, which is less than that of free space ($\varepsilon_r < 1$), and is therefore at a lower velocity than the velocity in free space $c_0 = (\mu_0 \varepsilon_0 \varepsilon_r)^{1/2}$.

A helicon wave can be produced in a plasma in a static magnetic field (Figure 4.21) The helicon source consists of an insulating tube about 50 mm diameter and about 1.5 m long containing the plasma gas at low pressure [$1 - 10^{-3}$ Pa (7.52×10^{-3} to 75.2×10^{-6} Torr)], surrounded by a solenoid which is used to produce a static magnetic field of about 0.01–0.1 T. A shaped aerial (Figure 4.21) surrounds the tube. The aerial is normally fed with RF at 13.56 MHz at which the wavelength is 22 m in free space.

The aerial around the tube induces a magnetic field in the plasma in the tube and a corresponding electric field is induced around this magnetic field. Electrons

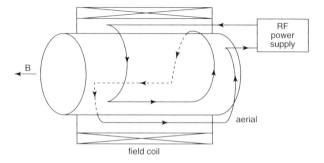

Figure 4.21 Schematic of a helicon reactor.

spiral around the magnetic field lines along the tube rather than across it, so that the electric field is greatest along the tube axis with the maximum field spaced by half a wavelength of the helicon wave, which since $\varepsilon_r < 1$ travels slower than the speed of light and has a shorter wavelength. The nondispersive acoustic wave is propagated in the direction of the diverging magnetic field.

References

1. Cobine, J.D. (1941) *Gaseous Conductors – Theory and Engineering Applications*, Dover Publications, New York.
2. Llewellyn-Jones, F. (1966) *The Glow Discharge*, Methuen, London.
3. Llewellyn-Jones, F. (1966) *Ionization and Breakdown in Gases*, Methuen, London.
4. Papoular, R. (1965) *Electrical Phenomena in Gases*, Illife, London.
5. Reed, T.B. (1968) Plasmas for high temperature chemistry, in *Advances in High Temperature Chemistry*, vol. **1** (ed. L. Eyring), Academic Press, New York, pp. 285–307.
6. Babat, G.I. (1947) Electrodeless discharges and some allied problems, electrical engineers-prt III: radion and communications engineering. *Journal of the Institution of Electrical engineers*, **94** (Pt. III), pp. 27–37.
7. Reed, T.B. (1961) Growth of refractory crystals using induction plasma torch. *Journal of Applied Physics*, **32** (12), pp. 25–34.
8. von Engel, A. (1983) *Electric Plasmas and Their Uses*, Taylor & Francis, London.
9. von Engel, A. (1965) *Ionised Gases*, Oxford University Press, Oxford (reprinted 1994, American Institute of Physics, New York).
10. Jewett, J. and Serway, R.A. (2008) *Physics for Scientists and Engineers with Modern Physics*, 7th edn, Thomson Higher Education, Belmont, CA.
11. Cambel, A.B. (1963) *Plasma Physics and Magneto Fluid Mechanics*, McGraw-Hill, New York.
12. Fridman, A. (2008) *Plasma Chemistry*, Cambridge University Press, Cambridge.
13. Brown, S.C. (1959) *Basic Data of Plasma Physics: the Fundamental Data on Electric Discharges in Gases*, MIT Press, Cambridge, MA (reprinted 1997, Classics in Vacuum Science and Technology, Springer, New York).
14. Roth, J.R. (1995) *Industrial Plasma Engineering, Vol. 1, Principles*, Institute of Physics, Bristol.
15. McTaggart, F.K. (1967) *Plasma Chemistry in Electrical Discharges*, Elsevier, Amsterdam.
16. Graham, W.G. (2007) The physics and chemistry of plasmas for processing textiles and other materials, in *Plasma Technology for Textiles* (ed. R. Shishoo), Woodhead Publishing, Abington, Cambridge, pp. 1–24.
17. Krall, N. and Trivelpiece, A. (1973) *Principles of Plasma Physics*, McGraw-Hill, New York.
18. MacDonald, A.D. (1966) *Microwave Breakdown in Gases*, John Wiley & Sons, Inc., New York.
19. Roussy, G. and Pearce, J.A. (1995) *Foundations and Industrial Applications of Microwaves and Radio Frequency Fields*, John Wiley & Sons, Ltd, Chichester.
20. Asmussen, J. (1990) Electron cyclotron resonance microwave discharges for etching and thin-film deposition, in *Handbook od Plasma Processing Technology*, (eds S.M. Rossenagel, J.J. Cuomo, and W.D. Westwood), Noyes Publications, New York.

Further Reading

Chen, F.F. (2007) *Advanced Plasma Technology*, Wiley-VCH Verlag GmbH, Weinheim, pp. 99–116.

Conrads, H. and Schmidt, M. (2001) Plasma generation and plasma sources. *Plasma Sources Sciences and Technology*, **9**, 441–454.

Raizer, Y.P., Schneider, M.N. and Yatsenko, N.A. (1995) *Radio-Frequency Capacitive Discharges*, CRC Press, Boca Raton, FL.

Rossenagel, S.M., Cuomo, J.J. and Westwood, W.D. (1990) *Handbook of Plasma Processing Technology*, Noyes Publications, New York.

Schmidt, M. and Conrads, H. (2001) Plasma sources, in *Low Temperature Plasma Physics* (eds R. Hippler, S. Pfau, M. Schmidt and K.H. Schoenbach), Wiley-VCH Verlag GmbH, Weinheim, pp. 283–304.

5
Applications of Nonequilibrium Cold Low-pressure Discharges and Plasmas

5.1
Introduction

At low gas pressures, charge equilibrium can exist but the separation of neutral gas molecules is such that insufficient collisions take place for thermal equilibrium in a plasma. Coupling energy into the plasma is difficult owing to the reduced number of collisions at low pressures requiring long residence times. At very low pressures, for example in electron and ion beams, charge equilibrium no longer exists and, since the residence time is short, high voltages or currents are necessary.

Some of characteristic properties of low-pressure glow discharges are listed in Table 5.1.

One of the difficulties of using plasmas at low pressures is coupling sufficient energy into the plasma. Methods used include inductively coupled plasmas (ICPs), helical and helicon sources and capacitively (E field) plasmas and power density enhancement using crossed electric and magnetic fields, using magnetrons, electron cyclotron resonance (ECR) and helical coupling (helicons) (see Chapter 4).

5.2
Plasma Processes Used in Electronics Fabrication

Low-pressure glow discharge plasma processes are used extensively in the manufacture of semiconductors and specifically computer memory chips. Some of these processes and the energies required are shown in Figure 5.1.

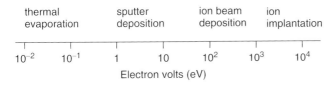

Figure 5.1 Ion energy required in different low-pressure processes.

Introduction to Plasma Technology: Science, Engineering and Applications. John Harry
Copyright © 2010 WILEY-VCH Verlag GmbH & Co. KGaA, Weinheim
ISBN: 978-3-527-32763-8

5 Applications of Nonequilibrium Cold Low-pressure Discharges and Plasmas

Table 5.1 Some characteristic properties of low-pressure glow discharges.

Mean free path at 1 Torr	$\sim 50 \times 10^{-6}$ m
Debye length	$\sim 10^{-4}$ m
Electron densities	$10^{15}-10^{18}$ m
Degree of ionization	$10^{-6}-10^{-4}$
Electron energies	1–8 eV
Ion energies	>10 eV
Gas pressure	0.1–1 Torr
Frequency	DC 13.56 MHz

Silicon-based integrated circuits today have up to 10^9 transistors on a single computer chip. Etched trench widths (the separation between components on a chip) are less than 0.2 μm wide. Manufacture involves a series of up to 12 different steps in which low-pressure gas discharge plasma processes are used. Some examples of these are shown in Figure 5.2.

The different processes are as follows:

1) film deposition (including alloy and organic compounds)
2) plasma-enhanced chemical vapour deposition (PECVD) at low substrate temperatures
3) ion implantation
4) deposition of masks
5) removal of masks (ashing)
6) high-definition etch patterns
7) nonreactive and reactive sputtering.

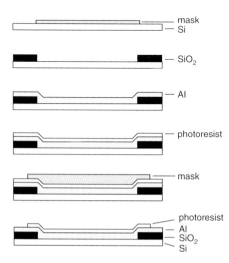

Figure 5.2 Different stages using plasmas in the fabrication of semiconductors.

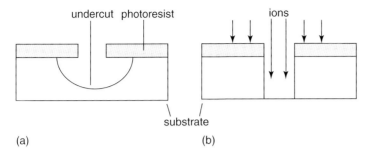

(a) (b)

Figure 5.3 Etching by (a) chemical process and (b) plasma.

Many of the processes require high definition of the etched or deposited layer. The ability to etch anisotropically with a high aspect ratio and high definition without the undercut that occurs with chemical processes (Figure 5.3) is one of the most demanding applications in semiconductor manufacture.

A typical reactive ion etching configuration comprises the wafer held in an insulated chuck at the bottom of a vacuum chamber. The etching gas such as sulfur hexafluoride flows through the chamber at a pressure up to a few hundred Pa (millitorr).

Some of the different processes and the gas pressures used are shown in Figure 5.4. The gas pressure is typically in the range $10-10^3$ Pa ($75.2 \times 10^{-3}-7.52$ Torr), at which the mean free path is between 5×10^{-3} and 5×10^{-5} m depending on the gas. Wall recombination of charged particles is important at low pressure, whereas at higher pressure particle collisions dominate. Operation at these low pressures presents the challenge of how to transfer energy from electrons by collisions to the atoms and ions required in the different processes.

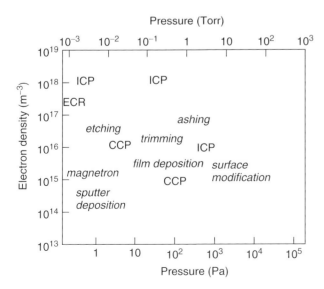

Figure 5.4 Examples of processes using low-pressure plasmas in the manufacture of semiconductors.

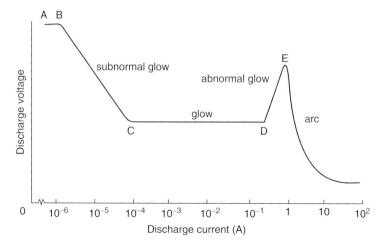

Figure 5.5 The generalized electric discharge characteristic showing the glow region.

5.2.1
The Glow Discharge Diode

Low-pressure plasmas generally have characteristics associated with glow discharges over the normal and abnormal region C–E of the discharge characteristic (Figure 5.5) where the current and power is highest before an arc develops and have high particle energies, but are normally in charge equilibrium.

The diode reactor [1] is a versatile plasma tool used in a variety of formats in semiconductor manufacture. Three electrode configurations are shown in Figure 5.6.

Glow discharges (see Secion 4.2) permit higher deposition rates than electron or ion beam processes and are used for thin-film deposition by sputtering from a target (cathode) on to a substrate (anode), or etching by sputtering from the substrate. A limitation is that the ion flux cannot be controlled independently of the ion energy (Tables 5.2 and 5.3).

The ion flux (number density) and ion energy, and hence the rate of deposition, can be controlled independently by adding a third electrode in the triode (Figure 5.6c). Increasing the negative bias increases the rate of deposition limited by onset of secondary ion emission at the third electrode and positive bias can be used to reduce the deposition rate. The electric field at the third electrode can be used to improve the directionality of the sputtered ions so that they arrive mainly perpendicular to the substrate surface and the undercut of the etched trench is reduced.

A typical operating pressure of a DC plasma diode (Figure 5.6a) is at a discharge voltage from about 500 V to 1 kV at up to 0.5 A and a current density of 10 mA mm^{-1}, with an electrode separation of 50 mm. The rate of bombardment depends on the current density and hence the size of the electrode. The space-charge limited

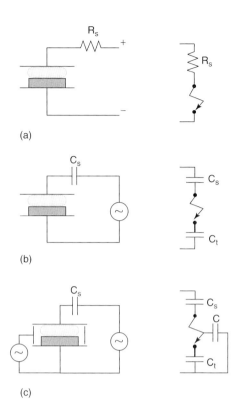

Figure 5.6 Diode and triode plasma reactor electrode and supply configurations: (a) direct coupled discharge; (b) AC capacitive coupling; (c) indirectly coupled triode.

current density (section 4.2.1.1) is

$$J_{sat} = A_0 T^2 \exp\left(\frac{-\phi e}{kT}\right) (A\,m^{-2})$$

The useful range of operating pressure of glow discharges is a compromise between the pressure below which the collision probability is too low for sufficient ionization and high enough deposition rates [<100 Pa (0.752 Torr)] and at higher pressures [>500 Pa (3.26 Torr)] at which nonionizing collisions, plasma instabilities and overheating of the substrate occur. The corresponding range of charge density lies between 10^{15} and 10^{18} m^{-3}, electron energies are between 1 and 20 eV and ion energies are of the order of 1 keV.

The directly coupled glow discharge with electrodes immersed in the plasma (Figure 5.6a) is limited to conducting substrates and when an insulating material is deposited the build-up of an insulating layer on the substrate stops the process. Control of the process is limited since it not possible to control ion flux and ion energy separately, since both are effectively a function of the voltage.

Electrodes used for radiofrequency (RF) etching are usually asymmetric since the substrate is usually smaller than the other electrode. A difference in the space charge in front of the electrodes exists so that the saturation current depends on electrode polarity and some degree of rectification occurs, so that the RF current

Table 5.2 Typical parameters of plasma sources used in microelectronic fabrication processes.

Source	Gas pressure p (Torr)	Electron temperature T_e (eV)	Electron number density (n_e m^{-3})
DC glow	0.1–1	1–8	10^{15}–10^{18}
Capacitively coupled	0.05–1	2–4	10^{15}–10^{17}
Inductively coupled	10^{-4}–10^{-1}	1–10	10^{15}–5×10^{18}
Magnetron	0.001–0.2	2–4	10^{14}–10^{16}
Electron cyclotron resonance	10^{-5}–10^{-2}	5–15	10^{18}
Microwave	10^{-3}–5	5–15	10^{14}–10^{17}
Helicon	0.01–0.1010	5–15	10^{18}–10^{19}

After Ref. [1].

has a DC component. This results in increased ion bombardment and sputtering at the substrate.

Secondary emission occurs when electrons are produced by bombardment of a material by electrons, ions or other particles. The particles transfer some of their kinetic energy to neutral particles at the surface which emits secondary electrons. The ratio of the average number of secondary charged particles to the number of equivalent primary charges in collision with the surface is the secondary emission coefficient δ.

In the hollow-cathode, secondary emission from ions accelerated in the cathode fall region oscillate within the hollow cathode between two metal sides of the cathode (Figure 5.7), resulting in a cumulative increase in electron emission without a glow to arc transition. The cathode current density can be increased by up to about 10 times the normal cathode current density without a glow to arc transition occurring. The space charge above the cathode surface is effectively increased in this way by using a tubular cathode or parallel plates to create a hollow cathode. The diameter or separation between the plates is about twice the depth

Table 5.3 Typical operating parameters of semiconductor fabrication processes.

Plasma typegrill	Pressure (Torr)	Ion density (m^3)	Degree of ionization
Deposition/etching	<10	<10^{19}	10^{-6}
Reactive ion etching	10^{-2}–10^{-1}	10^{19}	10^{-6}–10^{-4}
Magnetron sputtering	10^{-3}	10^{20}	10^{-4}–10^{-2}
Electron cyclotron resonance	<10^{-4}–10^{-2}	10^{21}	<10^{-1}

From Ref. [2].

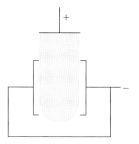

Figure 5.7 Hollow cathode.

of the negative glow region. The maximum current density varies with the depth of the cathode fall and hence the pressure and the wall separation.

5.2.2
The Magnetron

Magnetrons are used for the deposition of thin films. The limitation of the diode as an ion source is the low ion density at the low pressures [<100 Pa (0.752 Torr)] necessary in many electronic manufacturing processes. The long mean free path of the electrons results in few collisions with neutral atoms, which instead collide with the walls of the vessel.

In the magnetron (see Section 4.4.2.5), electrons are trapped in a transverse magnetic field produced by magnet pole pieces behind the cathode (Figure 5.8). Electrons leave the cathode in the vicinity of the north pole and move along the line of flux produced by the magnet, but require energy from the electric field perpendicular to the magnetic field to cross lines of flux.

The electrons rotate around the lines of flux at the gyro frequency $\omega_g = eB/m$ with a radius of typically 0.01–1 mm (correspondingly larger for ions), trapping them and further increasing their pathlength. The force due to the electric field eE accelerates the electrons in the direction of the electric field gradient and the electrons describe a spiral path of increasing radius as the flux density decreases, but decreasing velocity along the line of flux.

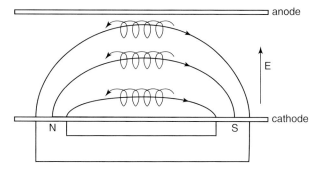

Figure 5.8 Motion of electrons in a magnetron.

Electrons leaving in the vicinity of the south pole will travel in the reverse direction. The absorbed energy is a function of the collision rate ν so that $\nu_c \ll \omega_g/2\pi$ to achieve high particle energies, that is, there need to be several revolutions by an electron before a collision occurs.

The discharge is maintained by ionization from secondary emission from the cathode due to ion bombardment. Pathlengths between electrodes of up to 300 mm are used and the increased pathlength results in plasma densities an order of magnitude higher than in the diode reactor. Operating voltages are of the order of 500 V at currents up to 5 A. The lower pressure reduces scattering and increases the energy of the sputtered atoms, which have a greater likelihood of the atom reaching the substrate. Other magnet configurations including circular and rectangular are also used. Deposition rates in excess of 10^{-10} m^3 s^{-1} are possible, but film microstructure, film purity and overheating the substrate occur at higher rates of deposition. Deposition rates up to two orders of magnitude faster can be achieved by bias sputtering.

RF magnetrons can be used for sputter deposition from insulating targets and for temperature sensitive targets such as plastics and metal oxide semiconductor (MOS) transistors. The rate of sputtering is usually less for RF than for DC magnetrons, since the fluctuating electric field leads to a less efficient confinement of the electrons and partly because electrically insulating targets often have reduced thermal conductivity and the energy input to the target is limited by thermal damage. Magnetrons are not widely used for etching due to their lack of directionality and also because of the higher heat transfer that occurs at the higher gas pressures used.

5.2.3
Inductively Coupled Plasmas

The induction plasma (Section 4.3.1) radio frequency solenoid coil (Figure 5.9) is a useful source of high-intensity plasma with the axis of the coil either vertical or horizontal for processes such as removal of photoresist, and etching in a barrel reactor and other noncritical applications and research where its low cost and flexibility are advantages. The induction pancake coil is useful for producing a

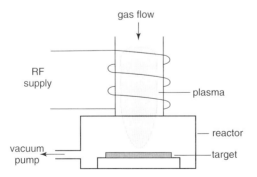

Figure 5.9 Induction plasma reactor.

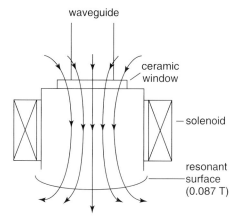

Figure 5.10 Electron cyclotron resonance reactor.

uniform plasma over a flat surface typically 100 mm in diameter at 13.56 MHz (Figure 4.10c).

5.2.4
Electron Cyclotron Resonance Reactor

Electron cyclotron resonance can be used to increase low charged particle densities and low energies at low pressures and is used for etching and plasma chemical vapour deposition (see Section 4.3.6) (PCVD).

An ECR reactor is shown schematically in Figure 5.10. For $\omega_{ce}\tau_c \gg 1$ and $\omega_g = mu_e/eB$, a very high absorption of power occurs at the resonant surface at the ECR frequency and can be coupled to a plasma by a waveguide and introduced into a vacuum chamber at about $0.133-1.33 \times 10^{-3}$ Pa ($10^{-3}-10^{-5}$ Torr) through a fused quartz window or with an aerial (the distributed electron cyclotron reactor (DECR) uses up to eight aerials). The microwave beam is right-hand circularly polarized (in the same direction as the gyro motion) or plane polarized where the left polarized wave has a field component at right-angles to it and does not contribute to heating the electrons.

The electric field strength in the plane of the waveguide is sufficient to ionize the gas and form a plasma in the reactor. The magnetic field along the axis of the reactor is provided by the solenoid. The diverging magnetic field accelerates the electrons rotating along spiral paths perpendicular to the magnetic field in the direction of the diverging field. A uniform magnetic flux density is produced over a plane of 0.0875 T (875 G) where the electric field is perpendicular to the external DC magnetic field at which the plasma resonant frequency is equal to the gyro frequency (see Chapter 4).

Electrons are heated in this region and pass on their energy to the slower moving ions. The slower moving ions are also accelerated in the same direction

Table 5.4 Comparison of properties of RF and ECR plasmas.

Parameter	ECR	RF
Plasma diameter (mm)	100	100
Plasma length (mm)	200	50
Frequency	2.45 GHz	13.56 MHz
Power (W)	500	200
Gas pressure (Pa)	0.0352 (0.4×10^{-3} Torr)	1.06 (75×10^{-3} Torr)
Electron temperature T_e (eV)	10	5
Electron number density n_e (electrons m^{-3})	10^{17}	10^{16}
Degree of ionization	10^1	10^{-2}

After Ref. [3].

and restrain the electrons by ambipolar diffusion so as to maintain the neutrality of the plasma.

The ECR discharge can be generated at pressures as low as of $0.133-1.33 \times 10^{-3}$ Pa ($10^{-3}-10^{-5}$ Torr) with charge densities the same as for RF glow discharges however the degree of ionization obtainable can be as high as 10%, at 10^{18} neutral particles per m^3 (0.01 Pa, 0.752×10^{-3} Torr) and is more than 1000 times that of an RF plasma operating in the range 10–500 Pa (0.752–37.6 Torr). A comparison of ECR and RF capacitively coupled plasmas is given in Table 5.4.

The uniformity and the degree of perpendicular path may be increased by using a magnet below the substrate to provide a flux parallel to the substrate surface and hence decreasing towards the substrate and reducing the angle of the electrons transverse to the axis at the substrate, which reduces undercut in the etching process.

5.2.5
The Helical Reactor

The helical reactor (Figure 5.11) uses a helical coil which resonates at a half or quarter wavelength of the supply frequency surrounded by an earthed coaxial cylinder forming a transmission line with a helical central conductor (see Section 4.3.4). Operation is at low pressure around 1.33 Pa (10^{-2} Torr). The helical coil behaves as a transmission line and circularly polarized electromagnetic waves are generated with the maximum radiation occurring along the helix axis. In the slow-wave helical resonator is a quarter or half wavelength mode resonant circuit. The velocity of the electric field is determined by the pitch of the helix and is a fraction of the speed of light. The field due to the current in the helix modulates the amplitude of the electrons increasing their energy along the length of the helix roughly as the square of the length along the helix axis.

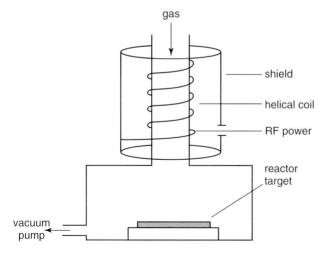

Figure 5.11 Helical resonator.

5.2.6
The Helicon Reactor

The helicon (Figure 5.12) supplies energy to an aerial outside the reactor inside a DC solenoid creating an electric field transverse to the magnetic field (see Section 4.3.4). A weak magnetic field from the source solenoid (0.005 T) couples the RF energy in the helicon mode into the centre of the plasma and some degree of confinement on the centre which diffuses into the lower chamber where it is confined by the second solenoid (0.007 T). Electrons tend to follow the ions which are accelerated in the diverging field along the axis of the coil to reduce charge imbalance. The acoustic longitudinal wave is propagated in the direction of the magnetic field and electrons in the plasma are directed along the flux lines and the ions follow by ambipolar diffusion, which acts as an inertia restraint to neutralize the charge separation.

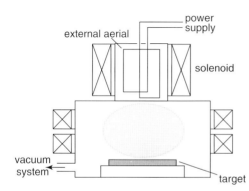

Figure 5.12 Schematic illustration of the helicon reactor.

5.3
Low-pressure Electric Discharge and Plasma Lamps

Electric discharges and plasmas emit part of their energy by recombination from excited or ionized states in the visible region of the spectrum (350–700 nm); however, most of the energy is radiated as thermal energy or by conduction and convection.

The hierarchy of different low-pressure glow discharge lamps is shown in Figure 5.13.

The efficacy (a measure of the efficiency of production of useful visible radiation of a lamp) depends on the availability of energy levels to permit suitable transitions for visible radiation. The availability of metastable levels, Penning ionization (see Chapter 2) and ease of ignition govern the choice of the gas, vapour or additives even in small quantities.

The colour of the plasma in the column of the discharge is characteristic of the gas (Table 5.5) and is sensitive to the type of gas, the discharge current and the presence of very small amounts of impurities.

5.3.1
The Low-pressure Mercury Vapour Lamp

Low-pressure mercury vapour discharges used in 'fluorescent lamps' use monatomic inert noble gases which have metastable excitation levels (see

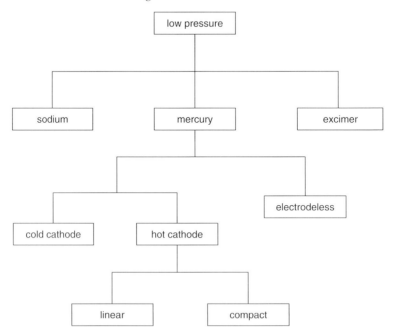

Figure 5.13 Different low-pressure discharge lamps.

Table 5.5 Characteristic colours for some gases and vapours in glow discharges.

Gas	First cathode layer	Negative glow	Positive column
Helium	Red	Green	Red to violet
Argon	Pink	Dark blue	Dark red
Neon	Yellow	Orange	Brick red
Krypton	—	Green	—
Xenon	—	Olive green	—
Air	Pink	Blue	Pink
Hydrogen	Brownish red	Pale blue	Pink
Nitrogen	Pink	Blue	Red
Oxygen	Red	Yellowish white	Pale yellow with pink centre

From Ref. [4].

Section 2.3.2.5) with low breakdown voltage. Operation is from AC, which prevents migration of positive ions to the cathode by electrophoresis and the electrodes are made from nickel to minimize sputtering.

Operation is over the normal/abnormal glow range of the discharge characteristic at currents from about 0.1 to 1 A and pressures of about 700 Pa (5.26 Torr) with electron temperatures of about 0.5–1 eV.

The hot cathode low-pressure mercury vapour lamp (Figure 5.14) uses a thermionic emitter (see Section 4.2.2) to achieve a low cathode fall voltage of about 15 V and a relatively high current density characteristic of a thermionic arc to achieve a higher efficiency than a cold cathode (Figure 5.15). The electrodes are made of tungsten wire that supports alkaline earth metal oxides such as barium strontium titanate, which, because it is highly sensitive to water vapour, is finally activated by passing current through the filaments at low pressure.

The internal tube diameter is less than the diameter of a free burning discharge and stabilizes the discharge and a build-up of charge on the insulated walls of the tube occurs due to collisions with high-velocity electrons that reach the walls. As a result, a negative voltage gradient exists close to the internal surface of the tube so that the discharge is well stabilized on the axis of the tube. The column does not constrict since the force due to its self-magnetic field is low and the effects of charge diffusion at about 700 Pa (5.26 Torr) are dominant. By decreasing the diameter of

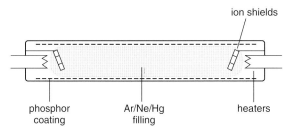

Figure 5.14 Hot cathode low-pressure mercury vapour discharge lamp.

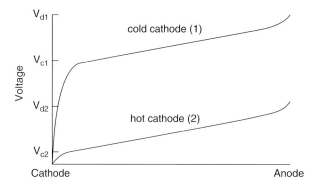

Figure 5.15 Variation of the voltage between electrodes of cold cathode glow discharge and thermionic arc cathodes.

the tube the column voltage gradient is increased, enabling compact lamps to be made.

The lamp contains argon with a small percentage of neon and is saturated with mercury vapour at about 1 kPa (7.52 Torr) vapour pressure. The neon limits the electron velocity by increasing ambipolar diffusion and reducing the loss of high-energy electrons to the walls, which would occur if only mercury was present, and lowers the breakdown voltage.

The colour of the positive column shows only line spectra from the gas, while that of the negative glow shows some of the electrode vapour lines (Table 5.5). The light output from the discharge is dominated by the spectral emission of the mercury. About 25% of the electrical energy is transferred via collisions between electrons and rare gas atoms and there is no excitation of rare gas lines. About 60% of the electrical input to the positive column is radiated at 250 nm (the resonance line of mercury), 15% at the 180 nm resonance line of mercury and 15% is distributed over other mercury resonance lines. The high-energy short-wavelength output from the mercury lines is used to excite lower energy transitions in the fluorescent coating on the inside of the tube at longer wavelengths in the visible region. Output in the near-ultraviolet is obtained by omitting the phosphor coating; however, the 250 and 180 nm lines are only transmitted if the borosilicate tube is replaced by quartz.

The coating on the filament has a high emissivity, which reduces the temperature of the filament so that a long filament lifetime is obtained. Ion shields modify the local potential distribution and attract the positive ions before they reach the cathode and protect the tungsten filaments from sputtering by positive ions. The majority of the current is carried by high-velocity electrons, which, because of their high velocity, miss the shields.

The cathode dark space in the arc mode is very narrow and the emission in the cathode region from the cathode material is different from the column (which is due to the mercury) and corresponds to the additive gas (argon).

Low-pressure actinic lamps are used as ultraviolet light sources for artificial sun tanning and photochemical processes, such as the manufacture of photoresist masks for printed circuits, erasing EPROMs, sterilization of water and for many

Table 5.6 Regions of the ultraviolet spectrum.

UV band	Wavelength (nm)	Photon energy (eV)
UV A	400–315	3.1–3.9
UV B	315–280	3.9–4.4
Actinic	320–200	3.9–6.2
UV C	280–100	4.4–12.4
Vacuum UV	200–100	6.2–12.4

applications in the printing area, including diazo printing and manufacture of offset lithographic plates.

5.3.2
Cold Cathode Low-pressure Lamps

Cold cathode low-pressure glow discharge lamps such as neon lamps are similar to the hot cathode mercury vapour lamp but the cold cathode has a cathode fall voltage of up to 400 V (see Section 4.2.2). The emission spectrum is from the gas (Table 5.6), typically at about 1–2.5 kPa (6–20 Torr), or a phosphor coating on the inside of the tube if mercury is used. Gases used include krypton, xenon and argon. The length of the lamp and the narrow bore result in a relatively high voltage gradient and lamps used for display and advertising often operate at voltages of 3–15 kV at currents of 20–120 milliamps.

5.3.3
Electrodeless Low-pressure Discharge Lamps

Electrodeless low pressure lamps are similar to low-pressure mercury discharge lamps but use a toroidal induction coupled (H field) glow discharge (see Section 4.3.1) operating at around 20 kHz. A primary winding is wound on a ferrite transformer core which is surrounded by the annular glass lamp (Figure 5.16). The toroidal discharge is the single-turn secondary.

5.4
Gas Lasers

The requirements for a plasma to excite a laser are not very different from those for a glow discharge lamp. However, the competing requirements of exciting the excitation wavelength matched to the laser transition often require more than one gas and a combination of metastable states and Penning ionization (see Section 2.3.2.5). Both glow and arc discharges are used in different lasers.

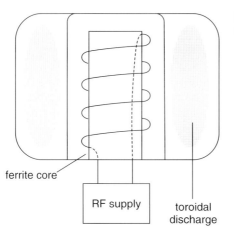

Figure 5.16 Example of electrodeless induction lamp.

The helium–neon laser is excited by electron atom transitions by the Penning effect in a glow discharge, a gas mixture ratio of helium to neon of about 20 : 1 at a pressure of about 1.3×10^3 Pa (10 Torr). Helium and neon have very close metastable levels, within about 0.05 eV. Helium is initially excited from the ground state to a long-lived metastable state which is close to the metastable level of neon. The excited helium atoms excite neon atoms in the ground state to the metastable level; any additional energy is supplied by kinetic energy. The neon atoms are then excited to the laser output level. The excitation transition is from the neon atom which is stimulated by closely coincident helium levels. Neon is poor at absorbing collision energy from the discharge but helium has a large cross-section. Energy transfer occurs from an excited He atom to ground-state Ne by an inelastic collision:

$$e_1 + He \rightarrow He^* + e_2$$
$$He^* + Ne \rightarrow Ne^* + He$$

The principal output is at 632.8 nm but other lines exist at 1150 and 3390 nm. Selection is obtained by maximizing the reflectivity at the required wavelength. Output powers in excess of 100 mW are obtainable but the main applications are for powers <5 mW at 632.8 nm.

A schematic arrangement of optical and electric field configurations used in gas lasers is shown in Figure 5.17. A high spatial stability is obtained by wall stabilization with a discharge tube a few millimetres in diameter and 100 mm to 1 m long depending on the power output. The bore of the tube does not extend the full length of the laser cavity and the cathode is not on the optical axis of the laser and can be made large enough for the cathode to operate in the normal glow region of the discharge characteristic. The root and fall regions of the discharge are kept outside the glow discharge, which reduces effects from sputtering of the cathode. The laser operates at about 1–1.5 kV at 5–100 mA. An ignition pulse of

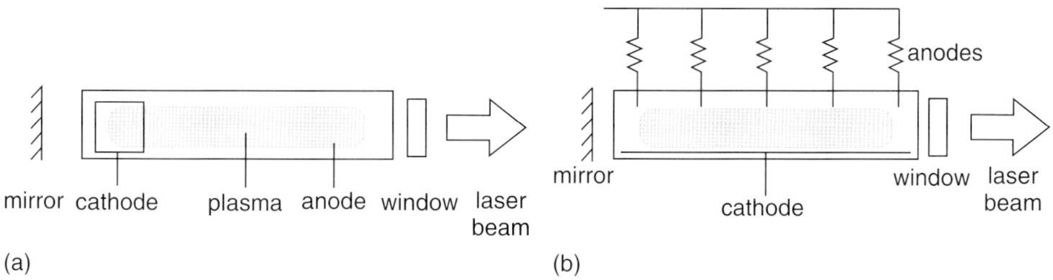

Figure 5.17 Electrode configurations in direct coupled gas lasers: (a) axial; (b) transverse.

about 8 kV is used to ignite the discharge. The efficiency of conversion of power from the electric discharge to light output is less than 0.1%.

Radiofrequency E field barrier (capacitive coupled) discharges are also used (Figure 5.18) to excite lasers over frequencies 50 kHz–100 MHz using external electrodes to avoid poisoning of the laser gas and degradation of the optics by sputtering.

The CO_2 laser uses an N_2–CO_2–He mixture at up to about 2 kPa (15 Torr). The nitrogen is excited to a lossless vibration level which enables a radiationless transfer from the metastable vibration level to the CO_2, and a population inversion is achieved. The return to the ground state takes place by collisions with He atoms with a laser output at 10 600 nm.

Excitation occurs over a narrow range of current and gas pressures in the normal glow regime. Laser transitions are quenched at temperatures above about 320 °C. Several different methods of construction are used depending on the power output, which is limited by cooling of the gas. Low-power lasers (<100 W) with a bore diameter of up to 10 mm use wall stabilization and the gas is cooled by conduction. At higher powers, large-area aluminium cathodes are used to minimize sputtering with pin anodes.

High-power CO_2 lasers use more than one discharge and cool the gas by recirculation either axially (fast flow) (Figure 5.18a) or transverse to the optic axis (Figure 5.18b). Discharge lengths up to 1 m (axial flow) or multiple parallel

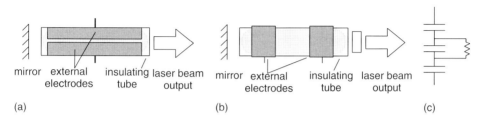

Figure 5.18 Electrodeless coupling used in gas lasers; (a,b) capacitive coupling, (c) equivalent circuit.

discharges (transverse flow) at discharge voltages up to 12 kV are convection cooled by passing the gas axially or transversely through the optical cavity.

Transverse excitation is used for molecular lasers; excimer lasers use a large number of separately stabilized pin anodes and a single plane cathode to achieve high current pulses without an arc developing. The efficiency of conversion is lower due to the relatively high fall voltage of the short discharge compared with the discharge voltage, which is only about 1 kV. The short current pulse suppresses arc formation and is also used with excimer or nitrogen lasers.

5.5
Free Electron and Ion Beams

The heart of electron and ion beam sources is the electron gun. Different designs are applied, using either thermionic emission (see Section 4.2.1.1) for which the electron output is from the cathode

$$J = A_0 T^2 \exp\left(\frac{-e\phi}{kT}\right)$$

or by field emission (Child's law) due to the high electric field in front of the cathode:

$$J = \frac{4}{9}\varepsilon_0 \left(\frac{2e}{mi}\right)^{\frac{1}{2}} \frac{V^{\frac{3}{2}}}{d^2}$$

Electron and ion beams are normally generated at very low pressures within an electron or ion beam gun over the range 0.01–0.001 Pa (0.0752×10^{-3}–7.52×10^{-6} Torr) at which the long free path enable high velocities and energies to be achieved without collisions, but are often used at pressures of 13.3 Pa (10^{-1} Torr) or higher. The electron beam can be deflected with a magnetic field so as to reduce evaporation on to the source.

5.5.1
Electron and Ion Beam Evaporation

Electron and ion beams are used to deposit thin films by evaporation and by sputtering of a target connected as anode (Figure 5.19). The electron beam is produced by thermionic emission from a cathode or by secondary emission at the cathode of a glow discharge in a vacuum chamber at low pressures of 1–0.01 Pa (7.52×10^{-3}–75.2×10^{-6} Torr); the mean free path ($4 < \lambda < 400$ m) is much greater than the separation between the source and substrate, and the electrons and the molecules evaporated from the source travel in straight paths without collisions.

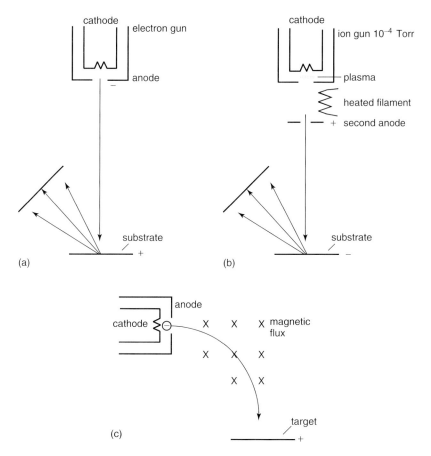

Figure 5.19 Electron and ion beam sources: (a) electron beam; (b) ion beam; (c) magnetic deflection of beam.

5.5.2
Ion Beam Processes

Ion beams are used for etching or sputtering by ion bombardment of a target connected as cathode (Figure 5.20) for cleaning the surface of semiconductor substrates and used in mass spectrometers.

The current in the beam is very low (microamperes), but ions are able to penetrate from a few nanometres up to about 1 μm. One method uses a glow discharge at a pressure above 0.001 Pa (7.52×10^{-6} Torr) as a source of high-velocity electrons, which in turn produce ions by collision. The ions are accelerated as a beam in a chamber at lower pressure. Diffusion of the beam due to charge repulsion can be reduced by electrons emitted from a hot filament at the outlet of the ion gun; which neutralize the high-velocity ions. Over the space-charge limited region ($J \propto V^{3/2}/d^2$) and for values of voltage of up to 120 kV (limited by production of X-rays and high voltage insulation required), ion current densities of about 10^{-5} A mm^{-2} at 1000 eV can be obtained.

When ion beams are used for etching, the energy and angle of incidence of the parallel beam of neutral particles can be controlled at the ion source independently of the processes occurring at the target and insulators can be sputtered without building up a surface charge. The thickness of sputtered film may vary from less than 1 nm to about 1 μm. The target may be sputtered over a wide angle but if the substrate is positioned in the flight path of the sputtered atoms or molecules most of them will be deposited on it. Sputtering uses ion energies from a few hundred to a few keV with corresponding voltages of 500–5000 V and above, for example in ion implantation. A separate electric field in front of the substrate can be used to accelerate the ions locally and, since neutral particles are not accelerated, impurities from the source are reduced. The ions reach the target without being diverted by collisions by evacuating the vacuum chamber to about 10^{-3} Pa (7.52×10^{-6} Torr).

Ion beam plating involves sputtering at pressures of around 0.1–10 Pa (0.752–7.52 Torr), at which much of the sputtered material is redeposited at the cathode and deposited by ions and neutrals from a separate source. Very high deposition rates are obtainable and the technique can be used both for repair and for coating components such as drill bits.

Ion implantation can be used to modify the surface of most materials to a depth of <0.5 μm at substrate temperatures <200 °C without the need for post-treatment or changes in surface finish. Commercial applications include manufacture of semiconductors, body implants and tooling for high-speed machining processes. Very high-energy focused beams of ions with a small range of velocities can be obtained. An ion beam implantation system is shown schematically in Figure 5.20. The material to be implanted is introduced in the ion source as a gas at pressures around 10^{-3} Pa (7.52×10^{-6} Torr) with a corresponding mean free path of 5×10^{-2} m and particle density up 10^{18} ions cm^{-3}.

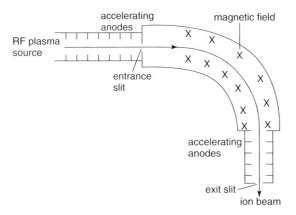

Figure 5.20 Ion implantation.

Positive ions are attracted from the ion source chamber by negative electrodes and pass through a magnetic field transverse to the direction of motion of the ions, which acts as a velocity monochromator. No deflection occurs when the force due to the electric field is balanced by the radial force from the magnetic field $Bue = Ee$:

$$u = \frac{E}{B}$$

and only ions of the same mass with a narrow range of kinetic energy will travel on the axis and through the exit orifice. The beam of ions is accelerated to energies of up to 10–500 keV, requiring accelerating voltages in excess of 100 kV, and are scanned over the target by an electric field.

5.5.3
High-power Electron Beams

Electron beam welding can be carried out in three different pressure regimes, hard vacuum 0.01 Pa (7.52×10^{-6} Torr); characterized by a stand-off from the work-piece of up to about 0.75 m, soft vacuum at about 13.3 Pa (0.1 Torr) and out of vacuum in air using a differentially pumped electron beam gun and a plasma window.

High-power electron beams are used for welding and melting with power densities of the order of 10^9 W mm^{-2} and powers of more than 300 kW at voltages between 60 and 200 kV at corresponding electron beam currents of 1 A to 40 mA. At low pressures, <1 Pa (7.52×10^{-3} Torr), the electron mean free path is of the order of 50 m so that few defocusing collisions occur and electron energies up to about 50 eV can be obtained. Heat transfer is from the high kinetic energy of the electrons accelerated in the electric field hitting the anode.

Figure 5.21 illustrates the main features of a high-power electron beam welder. The electron beam gun is evacuated to about 10^{-4} Pa (0.752×10^{-6} Torr). The thermionic cathode is either a directly heated thoriated tungsten filament or a tungsten disc heated by a secondary electron beam from a filament mounted behind it, enabling a robust cathode construction to be obtained.

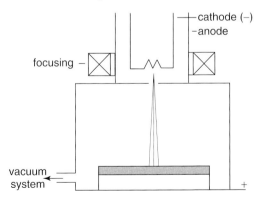

Figure 5.21 High-power electron beam for welding.

Electrons from the cathode are accelerated in the gun by intermediate anodes. Assuming no collisions, the velocity at the final anode is given by

$$u = \left(\frac{2eV}{m_e}\right)^{1/2} = \left(\frac{2 \times 1.6 \times 10^{-19} \text{ V}}{9.11 \times 10^{-31}}\right)^{1/2}$$

Putting $V = 10$ V gives $u_f = 5.9 \times 10^6$. In practice, some collisions occur, the number depending on the gas pressure, but over the range of pressures used similarity relations hold. Thermionic emission from the cathode is limited by the Richardson–Dushman equation (Eq 4.3). Below the space charge limited current density $J_{sat} I = pV^{3/2}$ where p is defined as the perveance.

The electron beam is focused by the axial magnetic field produced by the field coil, which changes the radial velocity components of the electrons to radial motion. Most of the electrons emitted by the cathode pass through the apertures in the centre of the intermediate anodes and are accelerated towards the work-piece, which is connected as the final anode.

The beam is scanned by two orthogonal pairs of field coils at angles. Accelerating voltages between the final anode and cathode vary from 20 to 150 kV with maximum power densities of the focused beam at the work-piece of more than 10^7 W m^{-2}. At voltages above about 30 kV, β-radiation (X-rays) is produced.

Typically, the maximum weld depth at 30 kV is 25 mm while at 150 kV it may be up to 1 m. Coupling of the energy in the focused beam to the vapour and plasma in the beam path and the local vapour pressure in the weld region may be up to 0.5 bar. Lower power densities can be used for many applications and it is possible to increase the chamber pressure up to about 10^{-1} mbar (10 Pa, 75.2×10^{-3} Torr), at which the mean free path is about 0.5 mm.

The electron beam can be emitted into the atmosphere over a distance of about 10 mm through a series of small orifices with differentially pumped chambers. The accelerating voltage required is increased to about 200 kV, β-radiation is produced and external shielding is necessary to protect the operators. If helium is used to shroud the outlet so that the electron mean free path is increased to about 0.186×10^{-6} at atmospheric pressure, the beam is defocused by collisions within a few millimetres from the output aperture. In-air electron beams have been used for continuous seam welding of tubes, cross-linking polymer films and sterilization.

At pressures above about 10^3 Pa (7.52 Torr), a cold cathode glow discharge can be used if the cathode area is sufficient to sustain a glow at the current required; the glow discharge requires stabilization. A shaped cathode glow discharge cathode can be used to produce a three-dimensional beam and it is possible to produce a single-shot circumferential weld around a tube using a ring-shaped cathode without rotating the beam or tube, or continuously heat-treat wire or a tube fed through a series of differentially pumped orifices.

The high efficiency of conversion of input electrical energy to thermal energy at the work-piece (>90%) due to the low energy loss from the beam at low pressures has led to the development of high-power electron beam furnaces for melting high-melting point metals. Melting is carried out in a water-cooled metal crucible

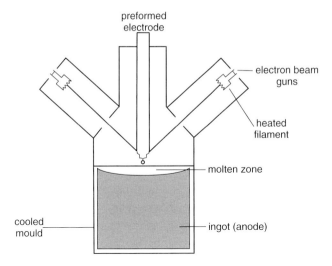

Figure 5.22 Electron beam melting furnace.

so as to minimize impurities from the crucible (Figure 5.22). Droplets of molten metal from the preformed electrode have a large ratio of surface area resulting in very effective degassing in the high vacuum. Electron beam furnaces are used with powers up to 1.2 MW using three electron beam guns operating at currents in excess of 200 A each.

5.6
Glow Discharge Surface Treatment

Plasma surface treatment by thermal diffusion is an important industrial process for the surface treatment of metals similar to the low-pressure DC diode reactor and uses the high electric field strength at the work-piece (cathode) to carry out surface reactions such as nitriding, carburizing, carbonitriding and boronizing. Ions from the gas at low pressure bombard the surface at gas pressures between 0.1 and 650 Pa. Some examples are given in Table 5.7.

The anode is the wall of the vacuum chamber (Figure 5.23).

The glow discharge treatment process for metals is illustrated schematically in Figure 5.23. The current is increased until the entire surface of the work-piece is covered. The maximum power density is obtained over the abnormal glow region of the discharge characteristic. The power density at the work-piece surface is given approximately by the product of the cathode fall voltage and the discharge current, $V_c \times I$. V_c is several hundred volts depending on the gas and gas pressure. DC electrode voltages of 400–1500 V at currents up to 300 A are used at current densities at the work-piece of 10–100 A m^{-2}, corresponding to power densities between 3.5 and 35 kW m^{-2}. Pulsed DC reduces the likelihood of a glow to arc transition occurring and a higher average current can be used. Operation in the abnormal

5 Applications of Nonequilibrium Cold Low-pressure Discharges and Plasmas

Table 5.7 Examples of low-energy plasma thermal diffusion treatment [5].

Process	Material	Plasma parameters
Nitriding	Steel	NH_3, N_2, $N_2 + H_2$ 0.1–1.3 Pa (8×10^{-4}–10^{-2} Torr) 673–873 K Maximum 40 h
	Cr-steel-TiC	85% N_2 + 15% H_2 1 kPa (7.5 Torr) Maximum 1023 K 6 h
	Ti alloy	1073 K 20 h
Boronizing	Fe	95% BC13 + 5% H_2 650 Pa (4.8 Torr) 1073 K
Carbonitriding	Steel	C-containing nitriding gas 983–1143 K 5 h
Carburizing	Steel Nb, W	1123–1273 K –

glow region increases the rate of treatment of the work-piece but also increases the possibility of the glow to arc transition. A glow to arc transition which results in a change in current density from 10 to 10^8 A m^{-2} in about 10^{-6} s is prevented from damaging components by rapidly acting solid-state switching circuits.

Local concentration of current density on a work-piece with an irregular surface creates a hollow-cathode effect (cross-reference) and results in a lower threshold of the glow to arc transition, but can be reduced by decreasing the gas pressure or current.

5.7
Propulsion in Space

The propulsion of a fuel-fired rocket relies on the expansion of a chemical propellant through a nozzle to provide the thrust. A plasma torch can be used to heat the propulsion gas rapidly without the need for a chemical reaction to achieve a similar effect. The thruster uses an arc jet similar to that of the plasma torch in which the plasma rapidly expands, increasing the axial velocity. The thrust can be increased using a magnetic field so that the plasma expands in the nozzle in the direction of the decreasing magnetic field and decreasing pressure (see Section 3.3). Some of the techniques used for semiconductor fabrication are also used for propulsion in space such as the as ECR and helicon reactors. This method of directing a low-pressure high-energy plasma has since been applied to PCVD, including the manufacture of diamonds.

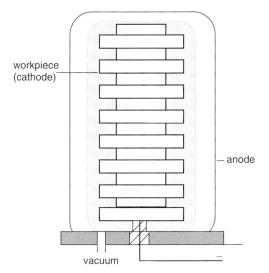

Figure 5.23 Surface treatment using glow discharge.

References

1. Roth, J.R. (1995) *Industrial Plasma Engineering, Vol. 1, Principles,* Institute of Physics, Bristol.
2. Das, A.C. (2004) *Space Plasmas: an Introduction,* Narosa, New Delhi.
3. von Engel, A. (1983) *Electric Plasmas and Their Uses,* Taylor & Francis, London.
4. Brown, S.C. (1959) *Basic Data of Plasma Physics: the Fundamental Data on Electric Discharges in Gases,* MIT Press, Cambridge, MA (reprinted 1997, Classics in Vacuum Science and Technology, Springer, New York).
5. Buecken, B. B. (1978) Erzeugung verschleissfester schichten in einer stromstarken glimmentladung. *Technik,* **33** (7), 395–399.

Further Reading

Overview

Boenig, H.V. (1988) *Fundamentals of Plasma Chemistry,* Technomic Publishing, Lancaster, PA.

Chapman, B. (1980) *Glow Discharge Processes,* John Wiley & Sons, Inc., New York.

Fridman, A. (2008) *Plasma Chemistry,* Cambridge University Press, Cambridge.

Fridman, A. and Kennedy, L.A. (2004) *Plasma Physics and Engineering,* Taylor & Francis, London.

Electronics Fabrication

Ahmed, N.A.G. (1987) *Ion Plating Technology,* John Wiley & Sons, Ltd, Chichester.

D'Agostino, R. (ed.) (1990) *Plasma Deposition, Treatment and Etching of Polymers,* Academic Press, New York.

Richter, H.A. and Wolff, A. (2001) *Plasma Etching in Microelectronics* in Hippler, R., Pfau, S., Schmidt, M. and Schoenbach, K.H. (eds) *Low Temperature Plasma Physics,* Wiley-VCH Verlag GmbH, Weinheim, pp. 433–452.

Rossenagel, S.M., Cuomo, J.J. and Westwood, W.D. (eds) (1990) *Handbook of Plasma Processing Technology,* Noyes Publications, New York.

Vossen, J.L. and Kern, W. (eds) (1978) *Thin Film Processes,* vol. **1**, Academic Press, New York.

Vossen, J.L. and Kern, W. (eds) (1991) *Thin Film Processes II*, Academic Press, New York.

Low-pressure Discharge Lamps

Lister, G.G. (2001) Low-pressure discharge light sources, in *Low Temperature Plasma Physics* (eds Hippler, R., Pfau, S., Schmidt, M. and Schoenbach, K.H.), Wiley-VCH Verlag GmbH, Weinheim, pp. 387–404.

Wharmby, D. (1993) Electrodeless lamps for lighting. *IEE Proceedings A*, **140** (6), 465–473.

Lasers

Hawkes, J. and Latimer, I. (1995) *Lasers, Theory and Practice*, Prentice-Hall, Hemel Hempstead.

Electron and Ion Beams

Brown, I.G. (ed.) (1989) *The Physics and Technology of Ion Sources*, John Wiley & Sons, Inc., New York.

Dearnaley, G., Freeman, J.H., Nelson, R.S. and Stephen, J. (1973) *Ion Implantation*, Elsevier, Amsterdam.

Dugdale, R.A. (1971) *Glow Discharge Material Processing*, M & M Monograph ME/5, Mills & Boon, London.

Harper, J.M.E. (1978) Ion beam deposition, in *Thin Film Processes* (eds J.L. Vossen and W. Kern), Academic Press, New York, pp. 175–208.

Powers, D.E. and Schumacher, B.W. (1989) Using the electron beam in air to weld conventionally produced sheet metal parts. *Welding Journal*, **68** (2), 48–53.

Quigley, M.B.C. (1986) High power density welding, in *The Physics of Welding*, 2nd edn (ed. J.F. Lancaster), Pergamon Press, Oxford, pp. 306–329.

Schultz, H. (1993) Electron beam welding, Abington Publishers, Cambridge.

6
Nonequilibrium Atmospheric Pressure Discharges and Plasmas

6.1
Introduction

Atmospheric pressure discharges (APDs) include glow, corona, barrier and surface discharges. APDs are not in charge or thermal equilibrium and can be used to carry out selective nonequilibrium chemical reactions with lower energy requirements without the need for vacuum equipment required at low pressures. The electron number density and electron energy are often at a higher mean level than for a plasma in charge equilibrium, and the discharge voltage and electric field strength are higher without the need for low-pressure facilities.

APDs have been studied for many years; major applications have been in electrostatic precipitators and for the manufacture of ozone for disinfecting drinking water and for depositing and removal of charge in the Xerox printing process. The principal limitation to their application has been the difficulty of producing a stable plasma at high voltages (>1 kV) at DC or power frequency at reasonable cost. The recent availability of low-cost high-frequency supplies using semiconductor inverters has led to extensive research in this area, including new applications such as surface treatment of textiles and medical applications.

Corona and barrier discharges are forms of pulsed discharges although they are not necessarily operated from pulsed power supplies but occur as rapid pulses by discharging local capacitance (barrier discharges) or where the field gradient is highest (corona). The presence of the conducting path modifies the field distribution and the local discharge is extinguished. The high current density and the short duration of the discharge prevent the discharge from establishing charge equilibrium and an arc developing, and enable high-energy nonthermal chemical reactions to take place.

6.2
Atmospheric Pressure Discharges

The discharge voltage current characteristic is shown schematically in Figure 6.1. The operating regime of corona discharges is shown schematically as the branch at B.

Introduction to Plasma Technology: Science, Engineering and Applications. John Harry
Copyright © 2010 WILEY-VCH Verlag GmbH & Co. KGaA, Weinheim
ISBN: 978-3-527-32763-8

6 Nonequilibrium Atmospheric Pressure Discharges and Plasmas

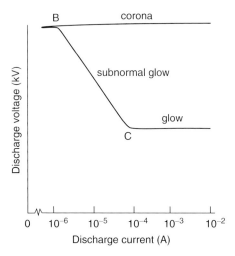

Figure 6.1 Non equilibrium region of operation of the nonequilibrium atmospheric pressure discharge.

Some examples of different nonequilibrium atmospheric discharges are shown in Table 6.1. The principal nonequilibrium discharges are corona and dielectric barrier discharges and the atmospheric pressure glow discharge. There are also subgroups such as microdischarges, partial discharges, barrier discharges and surface discharges.

The electric field strength of the atmospheric pressure corona discharge varies along its length and has high electron number densities of the order of 10^{14} electrons m^{-3}. The electron density of barrier discharges can be as high as 10^{16} m^{-3} at frequencies of 20–50 kHz, filaments diameters of 100 micrometers and current densities of 1 A mm^{-2} and nanosecond duration.

It is a characteristic of most of these discharges that they are pulsed, and the instantaneous current may be several tens of amperes. Pulsed discharges are dependent on the interaction of the electrical supply circuit and the discharge

Table 6.1 Properties of atmospheric pressure AC plasma.

Discharge	Discharge current (A)	Pulse length (ns)	Number of pulses per cycle	Gas temperature (K)	Pulse energy (mJ)
Dielectric barrier discharge	0.01–100	<100	>100	~300	<0.1
Corona discharge	0.01–10	~50	1–5	~300	~3.3
Atmospheric pressure glow discharge	0.01–0.20	~50	1–2	<1000	~2

(Chapter 10). Even fairly small values of inductance or capacitance may affect the current or cause resonance.

One of the difficulties in operating continuous glow DC or power frequency glow and corona discharges at atmospheric pressure, other than at very low currents, is preventing a glow to arc transition at currents above a few hundred milliamperes. If the discharge is supplied at a high frequency (>1 kHz), although the electrons have a high velocity and still traverse the gap, slower moving ions will oscillate at the supply frequency in the gap and the space charge is maintained at a value sufficient for secondary emission and allows operation at higher currents.

6.2.1
Corona Discharges

The Townsend region is characterized by high electric fields and low currents (microamperes) and enables high particle energies to be obtained. Although the mean free path in air at atmospheric pressure is less than 5×10^{-8} m, the electrons have sufficient energy to break chemical bonds and form free radicals in nonthermal chemical reactions.

The corona discharge is normally space-charge limited (Eq. 5.1) the voltage increases with current so that the discharge has a positive resistance characteristic (see Section 4.2.1.1). The mobility of ions is relatively low in the region beyond the ionized region and the discharge does not bridge the gap between the electrodes.

The corona discharge takes its name from the crown-like appearance of the gas flares of ionized gas surrounding the Sun that is seen during a total eclipse. Corona discharges were observed around the top of ships masts due to the discharge from charged clouds during electrical storms, and were known as *St Elmo's fire*.

The charged particles form a sheath of positive ions at a positive electrode, or electrons and negative ions at a negative electrode, since the corona discharge is normally space-charge limited (saturated). The electric field strength is much higher than the equivalent steady-state discharge at the same current. Although the mean free path in air at atmospheric pressure is less than 5×10^{-8} m, the electrons have sufficient energy to break chemical bonds and form free radicals and are therefore potentially suitable for carrying out nonthermal chemical reactions. In the corona discharge, the current increases with voltage so that the discharge has a positive resistance characteristic. The mobility of ions is relatively low in the region beyond the ionized region and the discharge does not bridge the gap due to divergence of the electric field, which varies approximately inversely with distance from the electrode.

A high electric field strength occurs at the surface of a conductor at high voltage which has a small radius, such as a point or wire, which is known as the *active electrode* and defines the polarity of the corona. If a sufficiently high voltage is available, electrons are accelerated and ionization occurs in front of the electrode due to the local increase in the electric field strength and a corona discharge is formed. The type of corona depends on the polarity of the electrode, the voltage at which spark breakdown occurs, the electrode geometry and surface

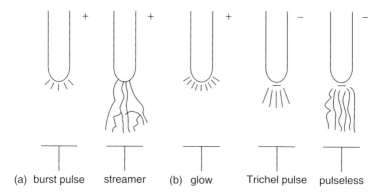

Figure 6.2 Various forms of corona discharges. (a) Positive electrode and (b) negative electrode.

condition (particularly asperities), current and whether the gas is electronegative or electropositive. Figure 6.2 shows examples of different forms of corona discharge for different polarities. The corona at a positive electrode starts as a noiseless glow up to a current determined by the electrode condition followed by a series of irregular pulses and streamers from the electrode. The streamers are acoustically noisy and emanate radio noise. As the current is increased, the streamers branch until a spark develops at higher currents.

The streamers are characterized by rapidly moving regions of higher luminosity and current density. Examples of glow and streamer corona discharges are illustrated in Figure 6.3, although the differences are not always well defined. The streamers in a corona discharge are similar to those that precede a lightning strike and other electric breakdowns and are regions of intense local ionization.

A negative electrode is characterized initially by pulses at frequencies of 1–100 kHz and up to about 150 µA with very fast rise times of the order of 1.5 ns, known as *Trichel pulses*. The Trichel pulses give way to a pulseless negative glow corona as the current increases, which in turn, depending on the surface of the electrode, develops into a tufted, rapidly moving and noisy corona before eventual spark breakdown. The spark breakdown voltage is higher than for a positive

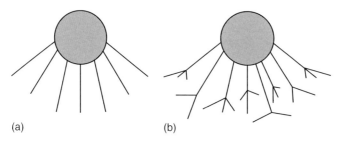

Figure 6.3 Glow (a) and streamer (b) corona discharges.

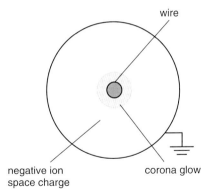

Figure 6.4 Corona between a wire and coaxial tube.

electrode and negative polarity is used where breakdown is to be avoided at high voltages such as in electrostatic precipitators.

The electric field strength in a corona discharge diverges approximately inversely with distance from the electrode until no ionization is apparent and the discharge ceases to exist. The current density at the electrode may be as high as that of an arc root; however, breakdown between the electrodes does not occur as the discharge does not cross the electrode gap. If only one electrode is used (unipolar discharge), the return path for the discharge current may not be apparent and is by the capacitive path to the supply; some energy is propagated as an electromagnetic wave, which, although not permitted today, was originally used for transmission of radio waves.

The electrical behaviour of coronas is illustrated by the electrode configuration commonly used in electrostatic precipitators of a wire radius a and a coaxial tube internal radius b. (Figure 6.4).

The voltage gradient between the tubes at a radius r length

$$E = \frac{q}{2\pi \varepsilon_0 r} \tag{6.1}$$

and since $V = \int_a^b E dr = \frac{q}{2\pi \varepsilon_0 \ln \frac{b}{a}}$ so $q = \frac{2\pi \varepsilon_0 V}{\ln \frac{b}{a}}$

$$E = \frac{v}{r \ln \left(\frac{b}{a}\right)} \tag{6.2}$$

showing that the maximum electric field strength occurs at $r = a$ at the surface of the wire. The conditions for corona can be determined in a similar way for wires, pointed electrodes and plane electrodes.

Positive ions, electrons and negative ions are produced by collision in the region close to the electrode, which are attracted by or repelled from the electrode depending on the electrode polarity. Electrons may become attached to electronegative gases such as oxygen to form negative ions. The charged particles form a sheath of positive ions at a positive electrode, or electrons and negative ions at a negative electrode, which when it is sufficiently conductive has the effect of increasing the effective size of the conductor and may limit the radial extent of the corona.

6.2.2
Corona Discharges on Conductors

A common configuration of high voltage conductors where corona may occur is two parallel conductors (Figure 6.5) such as those used in high-voltage power transmission or a wire to an earthed plane.

For two cylinders of radius a and separation d (Figure 6.5), by transposition of a conducting plane at the plane of symmetry at the midpoint between the cylinders (putting $E_d = E/2$) enables the maximum electric field at the cylinder surface to be found.

The electric field $d \gg a$, the electric field strength at the surface of the conductor is

$$E_d \approx \frac{V}{2a \ln\left(\dfrac{d}{a}\right)} \tag{6.3}$$

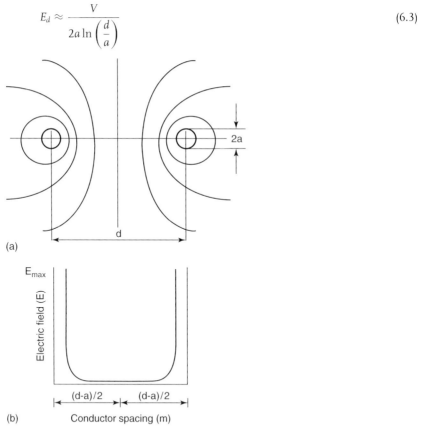

Figure 6.5 Electric field between two parallel radius a separation d wires ($d \gg a$). (a) equipotentials between electrodes, (b) electric field on axis between electrodes.

The discharge current is empirically determined as

$$I = k_m (V - V_0)^2 \qquad (6.4)$$

where V is the discharge voltage, V_0 is the inception voltage, I is the discharge current and k_m is a constant depending on the geometry.

The critical electric stress from Eq. (6.4) is given by

$$V_c' = m_0 \, g_0 \delta a \ln(d/a) \qquad (6.5)$$

which includes empirical coefficients for surface irregularity m_0 and barometric pressure δ_0 and the critical disruptive voltage.

We can define V_c' as the disruptive critical voltage to neutral at which ionization of the air immediately around the conductor commences but there is no visual corona, with a second voltage V_v' when the corona is visible. The critical disruptive voltage gradient is g_0, which at 25 °C and 1 bar is 2.18 kV.

The visual critical voltage (phase to neutral) is defined as

$$V_v = 2180 \, m_v \, r \delta (1 + 0.3 \sqrt{dr}) \ln \frac{d}{a} \qquad (6.6)$$

The power loss is given by Peek's equation, which is empirically derived as

$$P_c \left(\text{kW km}^{-1} \text{ per phase} \right) = 244 \times \frac{f + 25}{\delta} \times \sqrt{\frac{r}{d}} \times (V' - V_c')^2 \times 10^{-5} \qquad (6.7)$$

V_c is the minimum r.m.s. value of the voltage between phase and neutral at which the voltage gradient at the surface of the conductor reaches the voltage gradient required for breakdown.

The voltage at onset of corona is shown in Figure 6.6, corresponding to the inception voltage at the beginning of the subnormal region (Figure 6.1).

Corona discharges occur on high-voltage overhead power transmission lines and cause significant power loss, acoustic noise and radio interference. The corona may be caused by the proximity of an earthed pylon but the main source of power loss is due to corona between lines. The corona acts as a conducting sheath, increasing the effective diameter. The radial depth of the sheath can be determined approximately

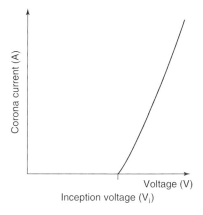

Figure 6.6 Onset of corona current at inception voltage between two parallel conductors.

from the radius at which E (Eq. 6.5) falls to the minimum value for electrical breakdown of air.

The corona sheath dissipates some of the energy due to voltage surges due to switching and induced by lightning. Fluctuations in the corona discharge result in oscillation at the natural frequency and harmonics of the supply circuit up to the microwave band, causing electrical interference. Water droplets act as capacitors and concentrate the corona discharge, causing acoustic noise as the water droplets fall.

6.3
Electrostatic Charging Processes

Insulating materials including fluids can be charged or discharged by passing them through an ionized gas, depending on its charge distribution. Air cleaning, particle collection, coating and printing are some of the areas where electrostatic charging processes are used.

Electrostatic charging processes use an electric field to create a space charge of free electrons and ions. The space charge may be used to charge either particles or a surface by attachment of the electrons or ions, which are then attracted to a charged surface as in printing or migrate to the surface of an electrodes as in an electrostatic precipitator.

A corona discharge can be used as a source of ions and particles can be charged by ion attachment. If the active electrode (where the electric field concentration is highest) is positive, it attracts negative charge, repels positive ions and acts as a source of positive ions.

Diffusion charging is due to the rapid thermal movement of ions in a space charge which results in attachment of ions to particles. Field charging is caused by bombardment by ions in a corona in an electric field and is normally the dominant process in corona charging and in diffusion charging.

When it is negative it attracts positive ions, repels negative ions and electrons, and acts as a source of electrons and negative ions. Charging occurs until the particle becomes saturated and further ions are repelled, and the magnitude of the charge is proportional to the surface area. Ionisation of the charged particle occurs in the electrode regions.

6.3.1
Electrostatic Precipitators

Electrostatic precipitators are used for the precipitation of dust from processes such as generation of electrical power and the manufacture of cement. Particle charging processes used in electrostatic precipitators operate over the Townsend and subnormal glow/corona regions B − B′ limited by electric breakdown of air at atmospheric pressure at about 3 kV mm^{-1} (Figure 6.1) corresponding to currents of the order of milliamps, and corona-glow discharges can be observed.

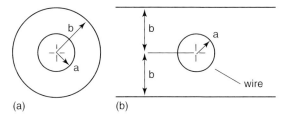

Figure 6.7 Tubular (a) and rectangular (b) electrostatic precipitators cells.

A common electrode configuration (Figure 6.7) is a coaxial central wire electrode of radius r_0 and a tube of internal radius R. Charging occurs close to the central electrode where the electric field is highest (Eq. (6.1)) and particles migrate towards the electrodes. The discharge varies with the electrode polarity and current, and whether the current is continuous or pulsed. AC is most frequently used but DC with a negative active electrode acting as the electron source is less prone to the development of sparks.

Particle charging occurs in the region between the boundary of the high-field region close to the inner electrode and the outer (passive) electrode. Corona initiation is greatest with the central electrode connected negative; the central electrode is usually made positive where corona formation is not the dominant process parameter so that the flashover voltage is increased. For particles of diameter above about 1 µm, charging occurs by ion attachment, field charging or ion impact. If the electrode is made negative, the electrode attracts positive ions and repels electrons and acts as an electron source. Depending on the ambient conditions, the electrons usually attach themselves to neutral molecules so as to form negative ions. Reversal of polarity results in positive ions being formed.

The high electric field close to the active electrode (normally negative) ionizes the gas to produce positive and negative ions and electrons. The larger particles (>1 µm) are charged by ion attachment by collision with charged particles from the corona in the inter electrode space. Particles of diameter <0.2 µm are charged by diffusion from the thermal motion of the ions and may occur independently of an electric field. The larger particles have higher saturation charge due to their size but lower migration velocities, and the saturation charge is much smaller than for the larger particles (Figure 6.8); however, the migration velocity increases as the diameter decreases further. The lower saturation charge density (Figure 6.8) of smaller particles is off-set by the different charging mechanism and lower mass induces charge on smaller particles.

The dust particles need sufficient time after they have been charged to travel to the electrodes, hence the length of the precipitator needed is a function of the gas flow rate. The time to charge the particles is very rapid, reaching around 90% of their saturation charge in less than 0.1 s, and is followed by the particle separation stage, which is at a lower electric field value and takes several seconds. Provided that the time of migration between the electrodes is less than the residence time τ_r, the particles will be removed.

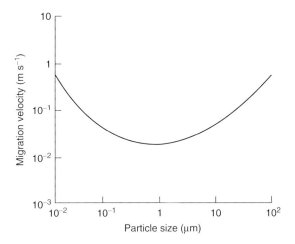

Figure 6.8 Variation of migration velocity with particle size.

For micron-sized particles and negligible turbulence, the migration velocity can be calculated by determining the balance of forces on a particle. If the charge per particle is q and the particle radius is r, then the radial force on a particle is qE_r, where E is the voltage gradient and the radial velocity is

$$u_r = \frac{neE_r}{6\pi\rho\eta} \tag{6.8}$$

where ρ is the particle radius and η is the viscosity.

The effectiveness is greatest with particle sizes above 10 μm, and the collection efficiency is proportional to the square of the electric field between the electrodes. The efficiency varies approximately exponentially with the length of the electrodes and the treatment time. A measure of the effectiveness can be determined to a first approximation from the collector geometry, $\eta = 1 - \exp(-A\omega/V_r)$, where A is the collecting electrode area, V_r is the volumetric flow rate and ω is the migration velocity.

For particles below 10 μm in diameter, a correction for Stokes law needs to be applied and the migration velocity becomes

$$u_r = \frac{2E^2rC_0}{\eta} \tag{6.9}$$

where η is the viscosity and C_0 is the Cunningham correction factor, which increases rapidly as the diameter becomes less than 1 μm.

Theoretically derived results show that the migration velocity decreases down to about 0.5 μm particle diameter, below which the migration velocity begins to increase rapidly due to the increased effect of diffusion charging at small particle diameters, which is consistent with measurements of collection rates.

Low-density insulating particles can be separated from conducting particles by feeding the mixture through a region containing positive and negative ions produced by a corona discharge.

The insulating particles become charged and adhere to surfaces of the opposite polarity but conducting particles do not. Electrostatic separators are used in areas such as mining, and for separating the shells and kernels of nuts.

6.3.2
Electrostatic Deposition

Electrostatic spraying can be used to spray coatings of charged materials in powder, liquid or fibre form in the range 10–120 μm. Particles of the material to be sprayed are entrained by a low-velocity air jet and are charged in the high-field region by collision between ions and electrons produced by the corona discharge in the space between the pointed electrode and the nozzle (Figure 6.9). The electrodes are at a voltage of 30–120 kV. A series impedance limits the current, and a high positive voltage gradient exists along the jet, which is normally operated about 25 mm from the surface. Particles are attracted to the work-piece, which is made positive. The particles lose their electrostatic charge on impact.

Film thicknesses from 10 to 200 μm are obtainable on cold surfaces which are subsequently fused thermally. Thicker films are obtainable by spraying on to a heated surface so that the particles melt and fuse. The local electric field is reduced as the coating builds up, resulting in a coating of uniform thickness. Similar techniques to those used for depositing coatings have been used for spraying liquids such as pesticides and provide a more uniform coating with a finer spray with less wastage and better coverage than conventional methods.

Corona discharges are used in photocopying and printing processes. A thin wire anode at a potential of about 10 kV produces positive ions which are attracted to the surface of a photoconducting material such as selenium. The image is exposed to light so that dark areas retain their charge and light areas are discharged so that negatively charged powdered pigment adheres only to the charged, unexposed

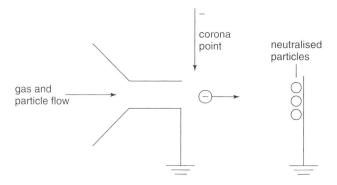

Figure 6.9 Electrostatic deposition process.

regions. The image is obtained by removing the uncharged particles mechanically and fusing the remaining charged particles to the paper. The charge remaining on the paper can be removed by a second corona source.

Build-up of electrostatic charge may also be reduced or eliminated by passing a charged material through a region in which the air is ionized, providing both positively and negatively charged particles in approximate charge equilibrium which combine with particles of opposite charge and neutralize the charge. Open-circuit electrode voltages of up to 24 kV are used at currents of about 0.5 mA, limited to a safe value by a high value of series impedance. DC is normally preferable to AC, since it is less affected by variation of the capacitance of the leads and interruption of the current at zero current. AC also has the advantage that the power source may be completely isolated by capacitive coupling through a dielectric such as silicone rubber to form a barrier discharge and is preferred for use in dusty atmospheres.

The ability of corona discharges to be attracted to areas of low impedance such as unevenness in coatings, defects or gaps is used for joint testing, pinholes and voids, uninsulated wires and for finding leaks in vacuum systems.

6.4
Dielectric Barrier Discharges

Barrier discharges with (a) one and (b) both electrodes insulated are shown with their corresponding equivalent circuits in Figure 6.10. Two forms of discharge are possible. The diffuse discharge is a single, uniform, homogeneous discharge similar to a glow discharge at low pressure, but can be operated at atmospheric pressure. The filamentary discharge consists of a large number of filamentary discharges at the dielectric surface similar to a corona discharge but the electric field does not diverge, resulting in a large number of parallel discharges.

Build-up of charge on the surface of the dielectric adjacent to the gap occurs until a series of microdischarge pulses characteristic of barrier discharges develop, and a single, uniform, diffuse, homogeneous discharge or a large number of parallel discharges at the dielectric surface.

If the discharge is supplied with a series of pulses of rapid rise time and short duration, the inertia of the heavier ions reduces their gain in energy, enabling the current to be further increased. The rapid interruption of the discharge prevents thermal equilibrium, and electron energies of the order of 1–10 eV are obtainable. The duration of the microdischarge is only a few nanoseconds and the transport of charge is less than 1 nC and is mainly due to electrons which have a higher mobility. At currents above the inception level, filamentary streamers approximately 100–200 µm in diameter originate from the dielectric surface where the electric field is highest. The resulting microdischarges or filaments in the discharge gap have a radius of only 0.1 mm and the discharge is uniform in the sense it is made up of a large number of filaments distributed over the whole electrode.

Dielectric barrier discharges, sometimes referred to as *silent discharges*, can be used to produce nonequilibrium atmospheric pressure plasmas. The discharge is

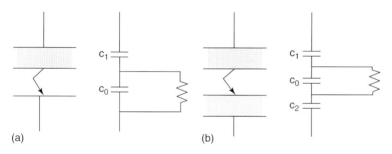

Figure 6.10 Equivalent circuits of different barrier discharge electrode configurations: (a) single layer and (b) double layer.

between the electrodes, one or both of which are covered by a dielectric layer. The current is stabilized by the capacitance of the barrier dielectric and distributed.

The effect of the low loss dielectric layer is to act as a distributed capacitive impedance which stabilizes the discharge between the electrodes by limiting the current density. The current is directly proportional to the electrode area and uniform high current discharges can be operated over a large area. If only one electrode is insulated (Figure 6.10a), the discharge length is limited to small separations by the discharges coalescing on the insulated electrode.

The barrier discharge is capacitively stabilized and the change in current as a result of a perturbation is

$$i = \frac{V}{R} e^{-\frac{t}{CR}} \quad (6.10)$$

If the time constant of the power supply output is progressively reduced, the stabilization decreases and eventually a series of discontinuous microdischarges occur.

The current is not limited provided that the electrodes are large enough to prevent the arc threshold current density from being exceeded. The discharge length is minimized to achieve the highest particle energies in the cathode fall region and a true plasma is not developed. Stable operation is obtained by using an RF power source typically at 20 kHz, at which the inertia of the ions is such that they cannot cross the electrode gap during a half cycle and the frequency is above the audio range.

The voltage across both the discharge gap and the dielectric in the absence of a discharge is given by $V_1/V_2 = C_2/C_1 = t_1\varepsilon_2/t_2\varepsilon_1$, where t is the spacing and ε_1 and ε_2 are the relative permittivities of the dielectric ($\varepsilon_1 = 1$).

In an RF electric field, with frequency ω, the displacement $x = eE_0/m_e\omega^2$.

Not all the charged particles will cross the gap and will be trapped in the gap between the electrodes, and the number will increase as the frequency increases. The drift velocity is very much greater for an electron than an ion, so the critical frequency is higher for an electron than an ion.

For a time τ taken for a charged particle to cross the gap t between the electrodes starting from zero velocity at one electrode, an angular frequency ω_c exists above which a charged particle starting from one electrode will not traverse the gap (see Chapter 3):

$$\omega_c = \frac{1}{2}\sqrt{\frac{eE_0}{2m_e d}} \tag{6.11}$$

The effect will depend on the relaxation time for the particle, which is very fast for an electron ($\sim 10^{-9}$–10^{-16} s); however, for the average number of electrons, although approximately equal to the number of positive ions, the lifetime is much shorter and the relaxation time of an ion (depending on the gas) is very much longer.

The frequency has a marked effect on the form of the plasma between the electrodes. Below the frequency at which ions are trapped, the plasma is difficult to initiate and forms a few aggressive filamentary discharges in the electrode gap. Above the ion critical frequency, a more uniform and diffuse plasma is formed, similar to a glow, while above the electron critical frequency a filamentary discharge is formed.

6.5
Plasma Display Panels

Plasma display screens use a similar layered electrode configuration to barrier discharges but with individual tiny cells between two layers of glass with the address matrix at the rear and the transparent display electrode and sustain electrodes at the front. Over a million cells may be used on a large screen. The current density is of the order of $1\ \text{A m}^{-2}$ and the individual cells are in the subnormal glow region, which is characterized by a high cathode fall voltage.

The panel is filled with a mixture of xenon and neon at about 53.2×10^3 Pa (400 Torr). Monochrome panels have a small amount of nitrogen added to improve the persistence. Colour panels use separate cells with red, green and blue phosphors. At the back of the panel are long electrodes behind a dielectric layer. At the front, the electrodes are perpendicular to the rear electrodes and are enclosed in an insulating transparent dielectric layer. An individual pixel can be addressed individually at the intersection of two perpendicular electrodes to produce a small discharge. A voltage of about 300 V is developed across the pixel, which drops by about half after breakdown and ionization.

6.6
Manufacture of Ozone

A simple medical ozonizer using coaxial electrodes with a dielectric barrier discharge is shown in Figure 6.11.

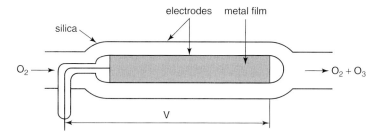

Figure 6.11 Ozone generator for medical use.

A major and well-established application of barrier discharges is for water treatment. Ozone is produced by a barrier discharge from oxygen or air, $3O_2 \rightarrow 2O_3$, but is unstable and it is necessary to generate it at the point of use.

Ozone is widely used for the large-scale disinfection of potable water and for swimming pools, odour removal and some therapeutic applications.

Ozone generation normally uses a battery of barrier discharges between coaxial tubular conductors (Figure 6.12) for which the electric field strength between the conductors is

$$E = \frac{V_{max}}{r \ln \dfrac{b}{a}} \qquad (6.12)$$

where b and a are the internal and external radii of the two tubular conductors. Very large ozone generators are made up of batteries of transparent tubes in which the ozone is generated and then passed through the water.

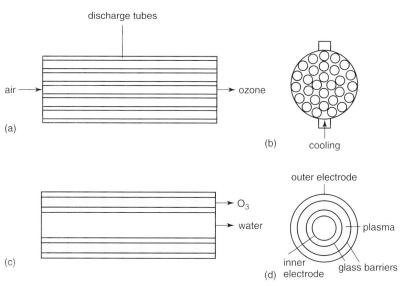

Figure 6.12 Battery of ozone generators for disinfection of water. (a) side (b) and end views assembly of discharge tubes, (c) side and (d) end views of a single discharge tube.

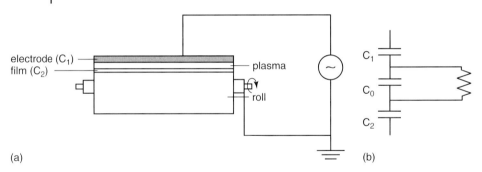

Figure 6.13 Plastic film treatment using a barrier discharge (a) discharge circuit arrangement, (b) equivalent circuit.

6.7
Surface Treatment Using Barrier Discharges

Barrier discharges are used for the surface treatment of plastic film used for packaging to increase the surface energy and improve hydrophobic properties and wetting and adhesive properties. The electrodes have a ceramic dielectric barrier and are supplied with a high-voltage AC supply at about 20 kHz. The plastic film covers the drive roll, which acts as the return electrode (Figure 6.13) and passes between the electrodes at high speeds prior to printing. The discharge voltage is around 1 kV but the high series capacitance and the requirement to break down the air gap with the insulating film in place requires an open-circuit voltage of about 20 kV with a correspondingly low power factor ($\sim$0.1).

6.8
Mercury-free Lamps

Recent concern about the accumulation of mercury from used lamps has led to the search for alternatives to mercury.

Some examples of different mercury-free lamps using barrier discharges at low pressures are shown in Figure 6.14. Barrier discharges in coaxial and rectangular discharge tubes using pulsed current to inhibit streamer formation using excimer (excited dimer) transitions (Xe_2^* at 172 nm and $XeCl^*$ at 308 nm) at up to 40% conversion efficiency are used to excite phosphors [Hipp]. Rectangular pulses are used to inhibit streamers at about 350 KPa(263 Torr). Flat panel lamps (Figure 6.15a) operate at 100 kHz, 30 W, 4.2 kV at overall efficiencies of 60%.

6.9
Partial Discharges

Partial discharges are discontinuous discharges in voids or inclusions in solid insulation or in bubbles in liquid insulation that partially bridge the insulation

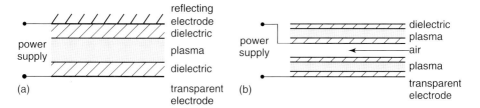

Figure 6.14 Mercury-free lamps. (a) flat panel and (b) coaxial lamps.

or at the boundary. The electric stress in the void is normally greater than in the insulation since the relative permittivity of the insulation is greater than the insulation and if the voltage is greater than the corona inception voltage (see Figure 6.6) a discharge occurs (Figure 6.15) which may continue until catastrophic damage eventually.

The equivalent circuit of the partial discharge is shown in Figure 6.14b as a series capacitor voltage divider with the discharge in parallel with one of the capacitors. The charge transfer associated with this type of discharge is often about 100 pC, typical of the streamer mode of the corona discharge, and a pulse discharge of a few nanoseconds to a microsecond duration occurs, before it extinguishes, after which the charge builds up again. The streamers result in damage to the surface and increase in conductivity, leading to a continuous AC glow discharge within the cavity over a long period and the void increases in size and the discharge increases in frequency. The discharge is of short rise time and duration and difficult to measure, but can be observed as a series of random current spikes.

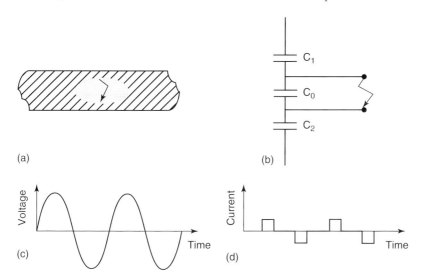

Figure 6.15 Partial discharge in insulation (a) discharge in cavity in insulation (b) equivalent circuit, (c) waveform of total current, (d) discharge current pulses.

The effect of the build-up of charge across the void and its subsequent discharge is to distort the local electric field and result in a small but discernible electromagnetic pulse; the small liberation of energy due to the capacitance discharge results in an acoustic wave and the discharge itself gives off light from excitation and ionization.

Partial discharges can also occur on the surface of insulation, for example of overhead lines where they cause pitting of the insulation.

The breakdown voltage of air (3 kV mm^{-1}) corresponds to 3 V μm^{-1}, so there is a limit to the minimum size of void at which a discharge will develop. Below this size the mean free path of air at atmospheric pressure (∼60 nm) is of the same size and discharge and ionization does not develop even under long exposure to an electric field.

6.10
Surface Discharges

Surface discharges are often a problem at high voltages and are the limiting voltage. The electric breakdown stress of air is reduced by moisture and dirt and breakdown of insulation occurs mainly due to moisture or contamination on the surface or at the joints.

Insulation at joints in high-voltage conductors, for example at the end of high-voltage transformer windings, feed-through insulation and at terminals are also often sources of unwanted breakdown, not only across the surface of the insulation but also between surfaces.

Below atmospheric pressure, the breakdown voltage across the surface of insulation decreases (Figure 3.9) due to the increased energy gained in particle collisions at low pressure.

A surface wave discharge can be produced at the interface of a plasma and a dielectric. The range of operating frequencies is from <1 MHz to over 40 GHz. The field breaks down over the surface between the electrodes (launcher). The plasma extinguishes when the plasma frequency $\omega_{pe} \approx \omega \left(1 + \varepsilon_g\right)^{1/2}$, where ω is the angular frequency of the supply and ε_g is the relative permittivity of the dielectric. Since the plasma frequency depends on the electron density and hence power, the size of the plasma also depends on the power.

The surface discharge (Figure 6.16) is another example of a discharge produced by a diverging electric field by using two electrodes of unequal size. The discharge

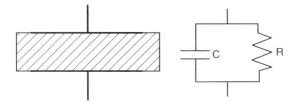

Figure 6.16 Surface discharges.

has similar characteristics to corona and barrier discharges. The electrodes are separated by a dielectric and the discharge travels over the dielectric surface between the electrodes. The voltage at the dielectric surface increases due to build-up of charge on the dielectric in the same way that charge builds up on the dielectric in the barrier discharge, reducing the voltage gradient and interrupting the discharge, and a series of microdischarge pulses occur in the same way as the barrier discharge.

Further Reading

Overview

Becker, K.H., Kogelschatz, U., Schoenbach, K.H. and Barker, R.J (eds) (2004) *Non-Equilibrium Air Plasmas at Atmospheric Pressure*, Taylor & Francis, London.

Bohm, J. (1982) *Electrostatic Precipitators*, Elsevier, Amsterdam.

Hippler, R., Pfau, S., Schmidt, M. and Schoenbach, K.H. (2001) *Low Temperature Plasma Physics*, Wiley-VCH Verlag GmbH, Weinheim.

Kogelschatz, U. (2003) Dielectric-barrier discharges: their history, discharge physics, and industrial applications. *Plasma Chemistry and Plasma Processing*, **23** (1), 1–46.

Parker, K. (2003) *Electrical Operation of Electrostatic Precipitators*, Institution of Electrical Engineers, London.

Ozone

Eliasson, M., Hirth, X. and Kogelschatz, U. (1987) Ozone synthesis from oxygen in dielectric barrier discharges. *Journal of Physics: Applied Physics*, **20**, 1421–1437.

Diesel Exhaust Treatment

Gundersen, M., Puchkarev, V., Kharlov, A., Roth, G., Yampolsky, J. and Erwin, D. (2001) Transient plasma-assisted diesel exhaust, in *Low Temperature Plasma Physics* (eds R. Hippler, S. Pfau, M. Schmidt and K.H. Schoenbach), Wiley-VCH Verlag GmbH, Weinheim, pp. 359–366.

Display Panels

Egitto, F.D. and Matienzo, L.J. (1994) Plasma modification of polymer surfaces for adhesion improvement. *IBM Journal of Research and Development*, **38**, 423–439.

Lee, J.K. and Verboncoeur, J.P. (2001) Plasma display panel, in *Low Temperature Plasma Physics* (eds R. Hippler, S. Pfau, M. Schmidt and K.H. Schoenbach), Wiley-VCH Verlag GmbH, Weinheim, pp. 367–385.

Shinoda, I. (2007) in Present status and future of color plasma displays, in *Advanced Plasma Technology* (eds R. D'Agostino, P. Favia, Y. Kawai and H. Ikegami), Wiley-VCH Verlag GmbH, Weinheim.

Shishoo, R. (ed.) (2007) *Plasma Technology for Textiles*, Woodhead Publishing, Abington, Cambridge.

Printing

Thourson, T.L. (1972) Xerographic development processes: a review. *IEEE Transactions on Electron Devices*, **ED-19**, 495–522.

Spraying

Kogoma, M. and Tanaka, K. (1990) Application of atmospheric pressure plasma: powder coating in atmospheric pressure glow plasma, in *Advanced Plasma Technology* (eds R. D'Agostino, P. Favia, Y. Kawai and H. Ikegami), Wiley-VCH Verlag GmbH, Weinheim, pp. 341–352.

Partial Discharges

Nattrass, D.A. (1988) Partial discharge measurement and interpretation. *IEEE Electrical Insulation Magazine*, **4** (3), 10–23.

Starr, W.T. and Steffens, H.G. (1988) Corona: insulation's silent enemy. *IEEE Electrical Insulation Magazine*, **4** (3), 41–46.

7
Plasmas in Charge and Thermal Equilibrium; Arc Processes

7.1
Introduction

Figure 7.1 shows plasmas in thermal and charge equilibrium, including arcs and inductively coupled plasmas. Applications generally use the thermal energy to carry out high-temperature thermal processes where high energy density and power are required.

Electric arcs are characterized by currents above about 1 A and temperatures of several thousand degrees celsius. The range of operation of the generalized characteristic is shown in Figure 7.1. The arc is normally in thermal equilibrium, and in the molecular gases dissociation and also ionization and excitation occur. The arc normally acts as a high-power heat source with a high temperature and enthalpy.

Some of the principal characteristics of arc process are as follows:

- low cathode and anode fall voltages (<25 V) adjacent to the electrodes but high voltage gradients in the fall regions
- high column temperature (>5000 K) and a relatively low voltage gradient compared with the fall voltages
- high current densities at the arc roots
- negative $V-I$ gradient of column.

Arcs are also characterized by evaporation at the electrodes, high current densities and self-magnetic fields, which cause constriction of the discharge and pinch and jet effects. Arc heating processes can be divided into use of three different regions of the arc (Figure 7.2).

Applications of arcs use either the electrode fall regions of the discharge, such as in welding where a short arc often only a few millimetres long is used, heat transfer from the arc column, such as the transferred arc plasma torch and the arc furnace, or both electrode and column regions.

Introduction to Plasma Technology: Science, Engineering and Applications. John Harry
Copyright © 2010 WILEY-VCH Verlag GmbH & Co. KGaA, Weinheim
ISBN: 978-3-527-32763-8

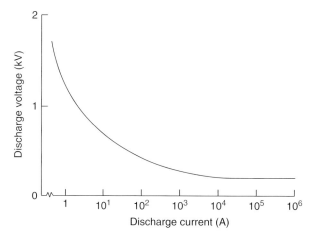

Figure 7.1 Part of the generalized characteristic corresponding to thermal arc plasmas.

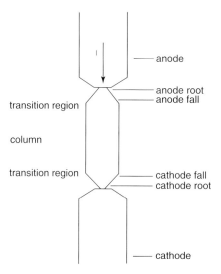

Figure 7.2 Principal regions of the high-current arc.

7.2
Arc Welding

Arc welding is a major industrial fabrication method and the different processes range from the apparently very simple AC or DC manual welding to complex high-frequency processes.

Modern power supplies use high-frequency inverters and electronic ignition and current control and are either AC or DC.

DC arc welding processes normally use an electrode connected as the cathode which also supplies the filler metal and the work-piece is connected as the anode. The arc stabilizes at the tip of the cathode where the temperature is highest and the pathlength is shortest, due to the effect of the self-magnetic field from the current in the electrode.

The power density at the cathode root is much higher than that at the anode. The electrode is normally a cold cathode material (m.p. <2000 °C) and melts at a rate governed by the heat losses from the cathode, including thermal conduction affected by the electrode diameter. The power density at the anode is normally lower than that at the cathode, but the power dissipated at the anode is higher, since the reduced constriction results in a higher total heat flux along the axis of the discharge. The column is short and the voltage drop small, and the power in the column is less than in the electrode regions and it is made as short as possible. The arc voltage V_d is equal to the sum of the cathode and anode fall voltages and the voltage drop in the arc column is approximately equal to the sum of the cathode and anode fall voltages, V_c and V_a, respectively:

$$V_d \approx V_c + V_a$$

Typical arc voltages are 15–20 V at arc currents of 10–200 A. A drooping output typical of manual welding is obtained by reducing the coupling of the supply transformer by saturation of the transformer core and forcing the flux between the windings to take longer paths in air due to saturation of the core.

The power input can be controlled by four main parameters:

- arc length
- arc voltage
- electrode feed rate
- arc current.

The arc voltage of the welding arc is directly related to the arc current and the current controls the heat flow. Manual welding processes generally use constant-current supplies as they are more tolerant of arc length and voltage fluctuations. Constant-voltage supplies can be used for automated welding such as metal inert gas (MIG) and submerged arc welding.

DC arcs are generally inherently stable and it is possible to operate at an open-circuit voltage as low as 60 V, as opposed to 80 V or more for an AC arc, due to the need for reignition of an AC arc every half cycle.

Deflection of the arc (*arc blow*) occurs, for example, during welding due to the self-magnetic field of the current in the work-piece, which has a perpendicular current component I_r resulting in a force $f(\overline{B} \times \overline{I_r})$ that drives the discharge along the work-piece away from the weld zone (arc blow). The magnetic force increases with the square of the arc current but can be overcome by balanced current connection (Figure 7.3).

Where AC is used (normally dictated by the process), a higher open-circuit voltage is required (100 V as opposed to 80 V) to assist in ignition and stabilization of the arc. The change in polarity each half cycle combined with different

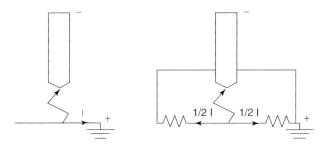

Figure 7.3 Effect of the self-magnetic field.

electrode materials and geometries may cause a variation in current amplitude on alternate half cycles and partial rectification and imbalance of the arc current. The AC voltage waveform has an ignition peak and an extinction transient. The plateau voltage between is approximately equal to the equivalent DC voltage. The current waveform is approximately sinusoidal with a short period corresponding to arc extinction and reignition each half cycle. Magnetic deflection of the arc on the work-piece is not a problem with AC since a steady magnetic field is not established.

The cathode may have a core or coating of fluxes, such as alkali metal oxides (MOs) or fluorides, which reduce the surface tension of the weld pool and which also have low ionization potentials, which assist in stabilizing the discharge.

A separate oscillatory magnetic field is produced in the same direction as the weld seam, and a stitching action is produced resulting in a wider weld bead; if the magnetic field is in the transverse direction to the direction of the weld seam, the arc root oscillates in the direction of the weld seam, resulting in a reduced seam width.

7.2.1
Metal Inert Gas Welding

The MIG welding process, introduced around the 1950s, virtually deskilled the up to then manual arc welding process. MIG welding uses a flux cored wire as a consumable electrode and filler. The wire is fed continuously over a sliding electrical contact through a guide nozzle with the shroud gas (normally carbon dioxide), which protects the weld zone from exposure to air and improves the weld quality and stability (Figure 7.4).

The wire feed rate can be varied to give different material deposition characteristics varying from dip transfer during which short-circuiting occurs at which the current is only just sufficient to melt the wire, to spray transfer where the wire is rapidly melted in small droplets. The change from dip transfer to spray transfer is accompanied by a change in arc voltage from about 20 V to more than 30 V. The arc length in spray transfer is longer than in manual arc welding and the arc voltage is higher so that the power in the arc column has an effect on the welding process. Different power supply characteristics are required for the different modes

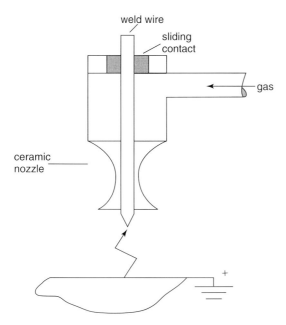

Figure 7.4 Metal inert gas (MIG) arc welding.

of operation. For dip transfer a constant current drooping characteristic is required, whereas for spray transfer the power supply has a relatively flat V–I characteristic, so that the process is less sensitive to variations in wire feed rates and arc length.

7.2.2
Tungsten Inert Gas Welding

The tungsten inert gas (TIG) welding process was developed mainly for welding aluminium and titanium protected from oxidation by an inert gas. A tungsten cathode is recessed by about 1 mm from the nozzle face and is protected from oxidation by an inert gas, normally argon, which passes through the insulating nozzle, maintaining the arc on the nozzle axis and also shielding the weld region (Figure 7.5). The weld filler material is added separately. The voltage gradient in the arc column using argon is lower than in air and the arc voltage is only about 25 V at arc currents from 10 to 500 A. Helium or nitrogen can also be used, allowing a higher arc voltage to be obtained. At higher currents, above about 120 A, a water-cooled cathode holder is used.

The ionization potentials of monatomic gases such as argon and helium are higher than those of diatomic gases, but the total energy required for ionization is generally lower in monatomic gases since no dissociation occurs (Figure 4.7). The current density is lower since the column temperature is lower (3000–5000 °C) and the column is wall stabilized by the ceramic torch nozzle. As a result, the arc in argon is more stable and quieter than in a diatomic gas and has a characteristic

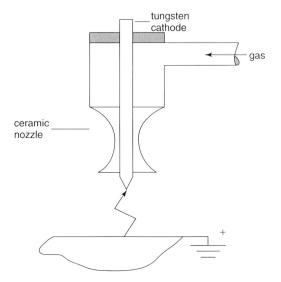

Figure 7.5 Tungsten inert gas (TIG) welding.

bell shape due to the divergence of the arc column in the gas flow leaving the nozzle.

The cathode is a thermionic emitter operated below the saturation current density ($J_{sat} \propto T^2$) (see Section 4.2.1.1). The shape of the cathode tip affects the local electric field, arc stability and heat transfer and hence the surface temperature. Pure tungsten can be used for the cathode for AC or DC welding; however, arc stability and arc ignition are improved by the addition of a lower work function material such as 1 or 2% thoria or zirconia for AC. Although the TIG arc can be ignited by touch starting if the electrode projects from the nozzle, this is not normally acceptable since it causes damage to the tungsten cathode and pick-up of tungsten in the weld and a high-frequency (>1 MHz), high-voltage (>3 kV) ignition supply is used.

With the tungsten electrode connected as cathode (straight polarity), about 60% of the heat transfer occurs at the work-piece. Reverse polarity with a copper anode in place of the tungsten cathode is used where tungsten impurities are unacceptable, for example in welding titanium, but at a lower current for the same electrode diameter. The arc voltage is up to 50% higher than for straight polarity but only about one-third of the arc energy is dissipated at the work-piece, resulting in shallow weld penetration and a wider heat-affected zone. Some applications use AC, which combines the advantages of both polarities. Ignition occurs more readily each half cycle with the cathode with a negative polarity, which can cause rectification where an imbalance of current occurs and the current is lower than when the electrode is positive. This can be overcome by using a power supply with a square-wave output or a separate ignition supply.

The TIG torch has also been used to fuse coatings and soften hard alloys prior to machining using a ceramic-tipped tool.

7.2.3
Submerged Arc Welding

Submerged arc welding is used for welding thick plates and for deposition of surface coatings. The welding electrode (cathode) provides the filler and the process is carried out at several hundred amperes underneath a layer of continuously fed powdered flux, which when molten protects the weld and provides a current path and which also stabilizes the discharge and protects the weld from air. Currents are as high as 2000 A using DC or AC.

Several arcs operated from the same supply (see Section 9.3.1) may be used for large welds or large surface areas using combinations of AC and DC supplies. Arcs connected in series have also been used to increase the heat input in submerged arc welding processes using DC. Repulsion occurs between series arc (anti-parallel) discharges, and single- and two-phase AC (in which the discharge currents are 90° out-of-phase) have been used to reduce arc interaction.

7.2.4
The Plasma Torch

The DC plasma torch used for fabrication of metals evolved from the Apollo space programme in the late 1950s, where it was used to spray refractory coatings on the nose cones of rockets and subsequently missiles. A plasma torch for profile cutting, welding and spraying is shown in Figure 7.6. The construction is similar to the TIG torch but the arc is constricted by the nozzle, which is electrically conducting and acts as an anode.

Constricting the arc column increases its voltage gradient and hence resistance and the power density and temperature in the column increase. The maximum temperature in the arc column can be increased from about 5000 K for a free-burning

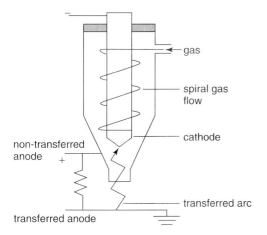

Figure 7.6 The DC plasma torch.

arc to 12 000–20 000 K, with a corresponding increase in enthalpy (total energy content). Molecular gases are used in cutting, welding and spraying, where rapid melting or vaporization of high-melting materials is required.

The separation between the cathode and nozzle is about 1 mm and requires a breakdown voltage of more than 3 kV, which is supplied by a separate high-voltage, high-frequency ignition supply in a similar way to a TIG arc.

In the non-transferred mode, the discharge is between the cathode and the nozzle and a jet of ionized gas in a relatively field-free region external to the torch is produced (hence the use of plasma). In the transferred mode, a resistance in the nozzle circuit limits the arc current to the nozzle to a value sufficient to maintain a pilot arc ($\sim$20 A). The plasma jet contacts the work-piece and the arc transfers to it and acts as an anode.

The plasma torch has application in specialized welding processes such as microwelding (<15 A), where the stability provided by the constricting nozzle and gas flow is better than that of a TIG, and the high-current welding of dissimilar materials. A greater tolerance to arc length and a longer arc and greater overall stability are also advantages. Examples of applications are welding high-melting or dissimilar metals where a high power density and localized heat input are needed, or where a high arc stability is required at low currents (<10 A), such as those used for the precision welding of metal bellows. Plasma torches are also used for cutting and welding under water where the gas jet provides a protective atmosphere around the arc, which is above atmospheric pressure.

Unlike other arc processes used for welding, the heat transfer from the constricted arc column is the main source of energy. The arc voltage $V = V_c + V_a + El$, however, while V_c and V_a are approximately the same as a free arc ($\sim$25 V), E is about 15 V mm^{-1} in air and diatomic gases. The *standoff* (separation between the nozzle face and work-piece) is a compromise between power in the column and obtaining a well-defined jet of high-velocity gas and is typically 8–10 mm, so that El is about 150 V and the voltage is as high as 180 V for profile cutting.

A high power density, spatial stabilization and constriction are essential for cutting and gouging, although these are less important in welding and spraying in which a larger nozzle orifice may be used where a lower arc voltage is acceptable. Argon, helium, nitrogen, hydrogen, carbon dioxide and argon–hydrogen mixtures can all be used with tungsten cathodes, but tungsten oxidizes in oxygen and cathodes using zirconium or hafnium which form protective electrically conducting oxides are necessary when air is used as the cutting gas. A separate shield of argon around the cathode has been used to protect it from the cutting gas which is added downstream. Water injection in the nozzle is also used to improve cut quality.

Plasma spraying is normally carried out in the non-transferred mode. The material to be sprayed is injected in the parallel section of the nozzle throat or at the outlet as a fine powder. The non-transferred arc is only a few millimetres long and the arc voltage is less than 50 V, so that high values of arc current up to 1000 A are needed to achieve adequate power input. The gases used depend on the material to be sprayed and include the monatomic gases and chemically inert and reducing gases. The process is affected by a wide range of parameters which influence the particle temperature and velocity and hence the adhesion, grain size and porosity of the coating.

7.3
Electric Arc Melting

Electric arc melting was one of the earliest applications of electrical power, ranging from melting a few grams of precious metals to remelting very large quantities of scrap and high-grade steel and special steel and iron alloys. Operation is at the supply frequency, in some cases above the normal supply voltage with inductive stabilization, although at the highest currents the self-inductance of the supply circuit may be sufficient. DC arc, plasma and hollow electrode furnaces are also in use. Very simple laboratory arc furnaces can be constructed around a refractory crucible using an arc welder as a power source.

7.3.1
The Three-phase AC Arc Furnace

The three-phase AC electric arc furnace (Figure 7.7) has been in use since the end of the nineteenth century for melting iron and steel alloys and other metals.

The three-phase electrodes are connected in delta configuration and held by electrode clamps above the furnace roof. Wear of electrodes is compensated for by slipping the electrodes in the clamps. The circuit is floating, that is, it is not connected to earth and the molten metal acts as a virtual earth. No separate

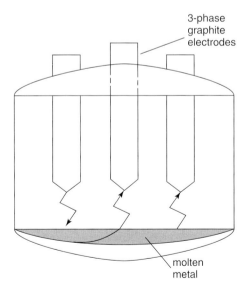

Figure 7.7 Three-phase AC arc melting furnace.

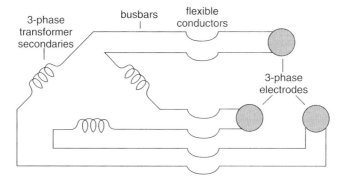

Figure 7.8 Low-reactance connections to three-phase arc melting furnace.

stabilization is shown since at the very high currents used the circuit impedance is sufficient. The delta connection is important at the high currents used (tens of thousands of amperes) to minimize self-inductance and skin effects. This is achieved by running the interphase connections of the secondary windings in parallel to minimize their self-inductance and connecting them at the electrodes, using parallel conductors (*busbars*) and flexible cables (Figure 7.8).

A typical arc furnace has a molten metal capacity of 110–130 tonnes with a furnace shell diameter of about 6.5 m and an electrical rating of 65 MVA. Open-circuit voltages of between 400 and 900 V AC and arc currents up to 70 kA are used.

The three arcs are separately stabilized and the molten bath provides a common electrode with three separate arc roots. A false star point (neutral) is obtained by a 'metal spider' buried in the refractories in the hearth as to give a neutral reference for control of the furnace. The minimum separation between the electrodes is governed by the attractive and repulsive forces ($F \propto I_1 I_2$) between the arcs which limit the current in the furnace. Three equispaced graphite electrodes of up to 0.7 m diameter about 1.6 m apart pass through the roof.

Each electrode is moved independently to control the arc voltage and current of the arc below it.

The length of the arc is longer than in many other arc processes and the voltage drop along the arc column is normally much greater than the sum of the electrode fall voltages:

$$El_{col} \gg V_c + V_a$$

Heat transfer to the metal during melting mainly occurs from the long arc column and during refining by radiation from the ends of the electrodes to the molten bath. The length of the arc during melting varies from about 250 mm during breakdown (when the side walls of the furnace are protected from direct radiation from the arc by unmolten scrap in the furnace) to short arcs of the order of 50 mm during refining and superheating.

The graphite electrodes behave as thermionic emitters when they are cathodes; however, the molten steel bath is at too low a temperature for thermionic emission and behaves as a cold cathode, and the high cathode current density causes high-velocity plasma jets at the surface of the bath. At the high currents used the arc voltage–current characteristic is almost horizontal or may even have a slightly positive characteristic due to the effects of plasma jets increasing the local voltage gradient and the self-electromagnetic field of the arc, both of which increase with the square of the arc current. The self-magnetic field results in a magnetic force which is high in the region of the arc root on the surface of the molten bath and causes rapid movement of the arc over the surface of the molten bath in the same way as a welding arc (arc blow).

The three-phase supply requires at least two arcs in series to complete the circuit. Three arcs form a series–parallel combination with a 120° phase angle between the current in each arc and the direction of electromagnetic forces between each arc varies between attraction and repulsion during the conduction cycle of the current. The self-magnetic fields of the electric discharges cause attraction (flow of current in parallel directions) or repulsion (anti-parallel). The force per unit length between two discharges of radius a and a distance x apart, $a \ll x$, is $F \propto I_1 I_2 \mu_0 / 2\pi x$, and if parallel discharges are sufficiently close they will coalesce with a shared column and separate attachment points at the electrodes.

The line current (in each electrode) is the vector sum of the current in the other two electrodes:

$$\bar{i}_1 = \bar{i}_2 + \bar{i}_3$$

and the line voltage measured between lines is equal to the sum of the arc voltages and the voltage drop across the stabilizing impedances. Differences in the current in each arc cause phase imbalance in the supply and the nonlinear arc characteristic causes harmonics on the supply. The very large load can also result in unwanted fluctuations in the supply voltage, causing interference and flicker of lights.

Each phase of the arc furnace can be represented separately by a fixed inductor in series with a variable resistor, corresponding to the arc and the resistance of the supply circuit. The locus of the voltage drop across X_L and the arc is a semi-circle of diameter $V = IZ$, in which $V\cos\varphi = IR$, $V\sin\varphi = IX_L$. Since X and V are constant, the vertical ordinate of the locus is proportional W and MVA and the abscissa is proportional to I. The circle diagram indicates that there are two operating conditions for the same value of power: one a long arc and high arc voltage, low current and high power factor, and the other a high current, short arc and lower power factor. The shorter arc is more stable and heat losses to the walls and roof are lower but the power factor is lower.

Rapid fluctuations in the arc voltage and current in each phase result in large variations in load current, generation of harmonics, current imbalance on the supply and flicker of lights where arc furnaces are connected to weak power transmission networks.

The arc current for a stable arc is approximately sinusoidal except for the region close to zero current since it is supplied through a series impedance of value comparable to the arc resistance, and because the arc is purely resistive it is in-phase with the arc voltage. The arc voltage (measured between an electrode and the bath) approximates to a square wave due to the flat V–I characteristic at high currents, except for the reignition peak and extinction transient.

7.3.2
DC Arc Furnaces

The construction of the DC arc furnace is similar to that of the three-phase arc melting furnace; however, small DC arc furnaces use only one arc since constant deflection of the arcs by the electromagnetic force between two closely spaced DC arcs would otherwise occur. This limits the arc current and furnace size. Furnaces with power ratings up to 48 MW and capacities up to 62 tonnes are in use and smaller ladle furnaces are used for refining steel alloys and for heating tun dishes to aid continuous pouring. Furnaces with higher power inputs and capacities are being planned. Electrical connection to the molten bath is obtained by using steel plates interleaved between the brickwork in the hearth or chromite bricks which have a relatively high electrical conductivity. Where the separation between the arcs is sufficient to reduce the force between them to an acceptable value, two electrodes are used, one of them serving as a current return path.

Figure 7.9 shows a DC plasma torch furnace of the type used for smelting exhaust catalysts to reclaim platinum group metals. The hollow electrode allows continuous feed into the centre of the zone heated by the arc. The hollow-electrode DC graphite electrode arc furnace is more stable than an AC arc furnace since reignition of the arc does not occur each half cycle. The more stable arc current results in a reduced loss of volatile alloy constituents and better utilization of the furnace and power source for the same average power input; however, the capital cost of the power supply is higher than that of the equivalent AC supply. The reduction in fluctuations of current results in lower electrode consumption, and also reduced flicker and acoustic noise. The current distribution in the electrode is uniform due

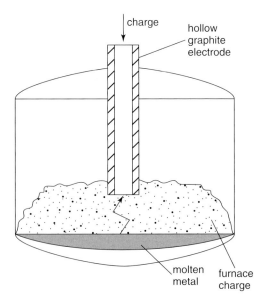

Figure 7.9 DC hollow-electrode arc furnace.

to the elimination of skin effects resulting in better use of the electrode and higher electrode current ratings.

7.3.3
Electric Arc Smelting

Three-phase electric furnaces are used for reducing ferrous and phosphorous ores and other high-temperature ($>1600\,°C$) endothermic processes in furnaces that are similar in construction to the arc furnace (Figure 7.10). The electrodes are buried in the charge, which consists of the mineral ore, flux and slag in which current flows by a combination of arcs and resistance in the charge.

The general form of the reduction equation for a MO with carbon is

$$2MO + C \rightarrow 2M + CO_2$$

which is normally highly endothermic.

Electrical conduction is by a combination of arcs and resistance through the furnace charge and the hot slag. The furnace is static and the operation continuous and the electrode arms are fixed. The Soderberg electrodes are formed *in situ* from carbon paste. Wear of the carbon electrodes is compensated for by slipping in the electrode clamps. Since the electrode arms are fixed, the load current can be balanced and the reactance of the supply to the electrodes minimized, for example by interleaving the conductors, and operation at higher currents than with arc furnaces is possible.

Larger diameter electrodes than those in arc melting furnaces can be used at currents up to 10^5 A, although the arc voltage is only about 100 V. The refractory

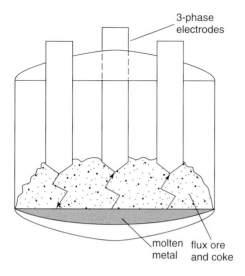

Figure 7.10 AC electric arc smelting furnace.

lining is protected from radiation from the arcs by the burden and long campaigns are possible before relining is required.

7.3.4
Plasma Melting Furnaces

Plasma melting furnaces use both DC and AC transferred arc plasma torches similar to those used for profile cutting and welding. The high temperature, enthalpy and spatial stability required for fabrication are not necessary for melting applications and the length of the nozzle throat can be reduced to a few millimetres and the nozzle orifice increased to 10 mm or more in diameter, resulting in reduced nozzle erosion and higher efficiency. Tungsten cathodes with flat ends can be used at up to 2000 A in argon, but rod cathodes with tapered ends are used at higher currents. Hollow-cathode torches using a copper cathode similar to those for heating gases, but used in the transferred mode with an external anode, avoid contamination by tungsten and have also been used for melting high-purity superalloys in cold crucible furnaces. The arc voltage is dependent on the separation between the cathode and molten bath surface but is typically 300–400 V at 5000 A at an arc length of 0.5 m. The maximum power rating is limited by the maximum cathode current, which is about 8000 A for a lifetime of about 200 h; as the arc current is increased, the operating life of the cathode decreases rapidly. The anode connections to the molten pool are similar to those used in the DC arc furnace.

A single torch mounted in the roof of the furnace is normally used, but three plasma torches mounted in the roof have been used for uniformly heating particulate materials fed through the coalesced region, preventing the build-up of unmelted material. A furnace with four torches mounted radially in the side walls so as to reduce arc interaction and heat the charge uniformly has been used for melting steel scrap.

High-power AC transferred arc torches have also been developed using a DC nontransferred discharge for stabilization. The use of three-phase AC has the advantages of acting as a balanced load and also eliminating the need for a bottom electrode, and prevents the development of stable interaction between arcs where more than one torch is used. Three torches operating at 50 Hz forming a balanced load at 2.8 MW (4000 A, 400 V per torch) have been used for melting steel, ladle furnaces and for refining and tundish heating.

Pre-reduced iron has been melted in particulate form in arc furnaces at up to 20 MW using hollow graphite electrodes and for reducing ferrochrome ores, fines and dust. The furnace is otherwise similar in design to the three-phase submerged arc furnace (Figure 7.10). The arc on the surface of the melt tends to repel the slag in the region of the electrode so that only a small quantity of fines is lost to the slag. Advantages claimed include the ability to process fines without the need for previous compaction, increased flexibility of the slag chemistry, reduction of unwanted reactions, improved alloy yields and lower electrode costs.

Other applications of DC plasma melting furnaces include a DC arc at about 200 V, 400 A between coaxial graphite electrodes immersed in a molten steel bath for superheating steel stabilized by argon or nitrogen between the electrodes; a radial discharge reactor in which a uniform region of ionized gas 150 mm in diameter is obtained from six coalesced DC arcs at up to 30 kW for in-flight plasma processes and as a heat source for a fluidized bed and a rotating plasma furnace.

7.3.5
Vacuum Arc Furnaces

The vacuum arc furnace (Figure 7.11) is used for remelting high-melting metals such as tungsten, molybdenum, titanium and zirconium which react exothermically with oxygen and are required with high purities where the ability to degas the metal and the absence of particles from crucible linings are essential. The preformed electrode is melted drop by drop in the vacuum vessel, allowing a high surface area for degassing and the surface tension at the surface of the molten pool tends to retain impurities.

The metal to be melted is connected as the anode. Arc voltages of about 30 V at currents up to 30 000 A DC are used. A DC field coil on the axis of the electrodes around the discharge is used to produce an axial magnet field of about 0.1 T. The axial magnetic field stabilizes the arc on the electrode axis by imposing a rotational velocity on any radial components of the arc current, preventing arc attachment to the sides of the water-cooled crucible. The droplets of molten metal have a high surface area, assisting degassing and impurities with a density less than that of the pure metal rise to the surface of the molten pool and solidify at the top of the ingot.

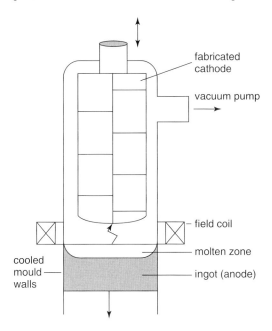

Figure 7.11 DC vacuum arc melting furnace.

Thermal stress causing cracks at either end of the ingot due to nonuniform cooling is also reduced.

7.4
Arc Gas Heaters

Arc heaters have been used for simulating space vehicles re-entering the Earth's orbit. Wind tunnels using high-velocity expansion of air through a nozzle result in cooling of the air which is heated by arc discharges at in hypersonic Mach 7 wind tunnels at powers above 10 MW. Methods used to produce the high-enthalpy gas flows include the use of magnetically rotated arcs and multiple electrodes. Many features of re-entry simulators, such as magnetic rotation of arcs, segmented electrodes and hollow cathodes, have subsequently been used in heaters developed for processing materials.

Different electrode configurations used for hollow-electrode arc gas heaters are shown in Figure 7.12. An axial magnetic field is used to rotate the arc and reduce electrode erosion. The operating voltage may be as high as several kilovolts, enabling operation at power levels of 1 MW to be achieved at arc currents of about 1000 A. Most arc gas heaters operate from DC power supplies; however, single-phase power frequency AC and three-phase arc heater configurations were developed for use in wind tunnels for re-entry simulation during the Apollo project in the 1960s.

In the nontransferred mode, the hollow tubular cathode is closed at one end with an open-ended tubular anode (Figure 7.12). The gas inlet is in the annular region between the electrodes and the arc is elongated by the effects of the gas flow and the electromagnetic interaction of the arc root on the electrode with its self-magnetic field due to the current in the electrode. The tangential gas inlet causes a gas vortex on the axis of the heater and stabilizes the arc on the axis of the reactor and also assists in rotating the arc. The speed of rotation can be as high as several thousand

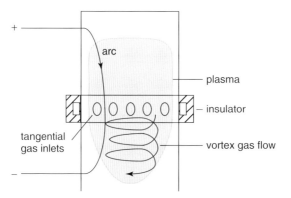

Figure 7.12 DC arc gas heaters.

revolutions per second. Electrically isolated sections in the anode can be used to reduce fluctuations in arc voltage and increase arc stability.

A three-phase arc reactor has been used at up to 1 MW for processing steel plant dusts fed into the arc column and a three-phase reactor was developed in the 1950s for the production of acetylene from coal gas. The superimposed mode of operation in which a three-phase AC discharge is maintained between the nozzles of cathodes of three torches stabilized by a DC nontransferred arc to form a large volume of plasma used for in-flight particulate reactions, gas heating and waste destruction.

A 1 MW plasma torch with a porous anode is in use for the dissociation of zircon sand to produce zirconia by injecting the sand into the long plasma tail flame of a three-phase arc.

- Electric Discharge Augmented Fuel Flames
- The maximum useful temperature of fuel–air flames is normally about 1500 K. Many industrial processes require temperatures between those of flames and arcs, in the range 1500–6000 K; however, as the temperature increases the gas enthalpy increases rapidly due to dissociation and ionization and the proportion of the electrical energy to the total energy input increases until at a temperature of about 4000 K the combustion energy is only about 10% of the total energy input.
- Thermal ionization in flames generates about 10^{14} electrons m^{-3} but a concentration of about 10^{16} electrons m^{-3} is required to establish a sufficient electrical conductivity to achieve a significant power input from the electrical discharge, which can be obtained by seeding with alkali metal salts to increase the degree of ionization. The tendency of the discharge to constrict, which results in a low-voltage, high-current discharge and nonuniform heating of the gas, can be overcome by turbulence in the flame.
- Electrically boosted burners combining 25 kW of combustion power and 10 kW (40%) of electrical power have been achieved. The temperature was claimed to increase from about 1800 to over 3000 °C with 90% conversion efficiency, excluding the heat losses in the exhaust.

7.4.1
Inductively Coupled Arc Discharges

The RF induction plasma torch uses an initial spark within the tube, for example from the high-voltage end of the induction coil; energy is coupled to the ionized gas from the power source and becomes self-sustaining. A toroidal plasma is produced with dimensions governed by the tube diameter and the skin depth. The swirling tangential gas flow tends to stabilize the plasma in the gas flow.

An RF ICP torch is shown in Figure 7.13. At the power levels and frequencies normally used (kW and MHz), resonant power supplies using electronic valves are used. RF inductively coupled (*H* field) discharges enable a clean source of ionized gas free from contamination from electrodes to be obtained and applications in spectroscopy are well established. The flow of feed is through the centre of the

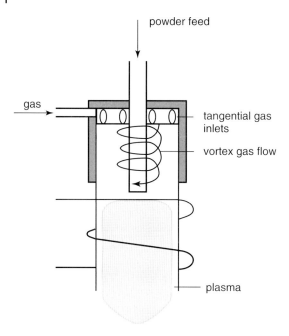

Figure 7.13 Induction plasma torch.

torch top. The residence time of feed materials is generally longer in an ICP torch than in an in-flight reactor using a DC plasma torch, because of the lower gas velocity and larger plasma volume.

The cross-section of the annular discharge is governed by the skin depth, $\delta = (\rho/\omega\mu_0\mu_r)^{1/2}$, where $\mu_r = 1$ and the radial depth of the conducting region is about 2δ. To achieve good coupling, $\delta \ll D/10$, some estimate of the skin depth can be made using characteristic values for an arc $E = 1\,\text{V}\,\text{mm}^{-1}$, $J = 10\,\text{A}\,\text{mm}^{-2}$,

$$\therefore \sigma = \frac{10 \times 10^6}{1 \times 10^3} = 10^4\,\text{S}\,\text{m}^{-1}$$

and the skin depth $\delta = 3.56$ mm at 10^6 Hz.

The power input to maintain the discharge increases as the frequency is reduced until at 10 kHz the minimum discharge diameter is more than 1 m and a power input of about 1 MW is required. Low-power torches such as those used for spectroscopy or laboratory applications (<30 kW) can be cooled by compressed air (<10 kW) or by water (<30 kW); however, higher powers for industrial applications require water-cooled ceramic tubes or axially segmented metal tubes which minimize eddy currents. Normal operating frequencies are about 6 MHz; although frequencies as low as 280 kHz at 20 kW in argon, with a discharge diameter of 65 mm and 1 MW at 450 kHz, with a discharge of between 75 and 150 mm using a water-cooled tube and frequencies up to 28 MHz, have been used. The overall efficiency of power generation using valve oscillators at high frequencies (>100 kHz) is less than 50%.

The discharge is initiated by the high E field between the ends of the coupling coil or to earth, which causes electrical breakdown. The H field is coupled into the discharge, which then becomes self-sustaining. The gas inlet is tangential and the radial pressure gradient due to the discharge current interacts with the high-frequency magnetic field, resulting in a magnetic force which increases with radius, causing a vortex which is superimposed on the axial gas flow within the discharge and improving the stability.

Gas-phase reaction processes have been carried out in ICP torches, although the yields were low, presumably due to bypass of the annular hot zone. Titanium dioxide has been produced by reacting heated oxygen with titanium tetrachloride and boron chloride has been reduced on an industrial scale to produce small quantities of boron metal of high purity and fine particle size.

Refractory materials can be spheroidized by feeding powder through an ICP; applications have included uranium pellets used in nuclear reactors, reflecting-silica beads and ink transfer particles used in photocopiers. Recent interest is in the production of nanosized particles.

Particle feeds cool the discharge and only relatively low feed rates are possible without extinguishing the discharge. There is only a small difference in the maximum and minimum radial temperature distributions at high frequencies, of the order of 7% at 4 MHz with an axial temperature of 10^4 K; however, the particle feed tends to bypass the hottest regions of the plasma, which have a higher viscosity than the cooler regions. By feeding the material into the tail flame of the discharge instead of the discharge itself, problems of instability, particle entrainment and feed rates are increased, but the heat transfer rates and residence times are reduced.

An interesting application is for the manufacture of fused quartz, where the absence of water vapour in the heating process eliminates absorption due to the OH radical at 2800 nm in the near-infrared.

Crystals of refractory materials have been produced by melting a continuous feed of powder in an ICP torch by the Verneuil method. Contamination from a crucible or combustion products is eliminated and it can be used when the electrical conductivity of the crystal material is too low for heating by induction.

7.5
High-pressure Discharge Lamps

High-pressure discharge lamps (Figure 7.14) radiate a continuum of wavelengths rather than the individual lines or bands of low-pressure discharge lamps and the spectrum is varied by changing the gas pressure and gas composition and a phosphor coating is not used to change the wavelength output.

The light obtained from a high-intensity discharge (HID) lamp is dependent on the gas composition, which is normally a vapour decomposed during warm-up of the lamp. Gases and vapours used include mercury vapour, metal halides, sodium vapour, sulfur vapour and xenon. An important factor is the colour rendering index (CRI), which is that of the vapour.

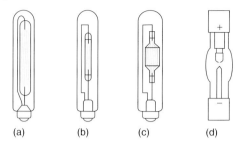

Figure 7.14 High intensity discharge lamps:
(a) low-pressure sodium vapour lamp; (b) high-pressure sodium lamp; (c) metal halide lamp, (d) high-intensity short arc lamp.

HID lamps generally operate with relatively short arcs at currents of around 1 A or more, at pressures close to atmospheric pressure and with wall loadings in excess of 3×10^6 W m^{-2}. The lamp voltage increases since as the current is increased, the pressure increases and hence the voltage gradient increases to a maximum and then decreases to an equilibrium value at point C (Figure 7.15). High-pressure lamps are normally easy to start from cold when the pressure is low, but reignition of a hot lamp at a high gas pressure is difficult.

Early high-pressure mercury discharge lamps used a simple tubular construction with a cold cathode operated below atmospheric pressure in mercury vapour, initially at low pressure with a light output in the blue and UV part of the spectrum, but have mainly been replaced by the more efficient high-pressure sodium lamp Figure 7.14a.

The low pressure mercury and sodium vapour lamps are similar in design but the mercury is replaced by sodium vapour, giving a characteristic yellow light (Figure 7.14a). As the temperature and pressure is increased, the light output occurs at shorter wavelengths. The principal developments have been due to improvements in materials, notably the alumina tube surrounding the discharge (Figure 7.14b), which becomes incandescent and re-radiates to give a light output closer to the spectrum of daylight.

The light spectrum in the visible region (350–750 nm) is increased further by operating an arc in an atmosphere of vaporized metal halides in argon with mercury

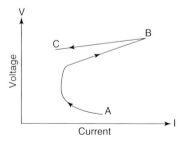

Figure 7.15 Voltage and current in a high-pressure discharge lamp during warm-up.

in a fused quartz tube (Figure 7.14c). The temperature profile is dominated by the ionization and radiation properties of the metal rather than the gas. The effect of the halide is to reduce the average excitation potential below a critical value $V' < 0.585$ V, enabling the arc discharge to operate in thermal equilibrium, not possible in the metal vapours alone due to their low vapour pressure. Metal halides such as ThI_4 have vapour pressures of 1 kPa (7.52 Torr) or more at temperatures within the range of operation of the quartz walls of the lamp. Initially the arc is relatively constricted in the mercury vapour, which forms a high ionization potential gas; the halide molecules which are initially condensed on the tube walls evaporate and dissociate into the arc, lowering the mean ionization potential, and the arc expands and becomes wall stabilized.

HID metal halide lamps are increasingly used for motor vehicles. The halide powder is heated and vaporized by the arc in an atmosphere of xenon. The xenon allows a more rapid warm-up than static HID lamps using argon. The gas pressure in the lamp bulb rises rapidly with an initial controlled current surge, permitting a rapid light response, and is stabilized at the normal operating voltage and current. An igniter and ignition electrode is used to ignite the arc even when hot at high pressure. The power supply uses an inverter to produce a square-wave output which is transformed to about 70 V AC. The lamp power is about 40 W and is at least 30% more efficient than the equivalent tungsten–halogen lamp.

A short-arc DC high-power arc lamp using xenon and mercury vapour is used for cinema projection and similar intensity focused light sources. The light arises from recombination of electrons in collision with ions in the space charge about 0.1 mm above the cathode surface in mercury vapour (Figure 7.14d). The operating voltage is about 40 V at several hundred amperes and more than 14 kW.

Recent concern about the accumulation of mercury from used lamps has led to the search for alternatives to mercury. *Excimer* (excited dimer) lamps such as the diatomic sulfur molecule (S_2) as a powder in argon excited by a microwave power source at 2.54 GHz with an output peak at 510 nm are nontoxic and are capable of efficient operation at several kilowatts at high efficacies, but require cooling and screening from emitting electromagnetic radiation.

The sulfur lamp uses dimer molecules (S_2) which produce light peaking at 520 nm. The molecular transition emission spectrum is continuous through the visible spectrum. The light is low in the infrared and ultraviolet regions of the spectrum with up to 73% in the visible part of the spectrum. The lamp is supplied at 2.45 GHz by a magnetron which is coupled via a hollow spindle which acts as a waveguide to a quartz sphere about 30 mm diameter containing sulphur and mercury vapour. (Figure 7.16). The gas breaks down and resonates at 2.45 GHz. The spindle is used to rotate the lamp to prevent localised over heating. The whole is enclosed in a Faraday cage to reduce EMI. The sulfur heated by the microwaves sublimes and increases the pressure to about 5 atm. Typical power inputs are 1.4 kW at an efficacy (light output normalized to eye response power input) of 100 lm W^{-1}.

High-power pulsed and continuous discharge lamps are used for the excitation of continuous YAG lasers. The lamps are water cooled and operate with DC discharges up to 200 mm long at up to 50 A in xenon or krypton at about 26.6 kPa (200 Torr). The highest light output over the region 200–1100 nm is obtained using xenon, which has a lamp efficiency (lamp light output/electrical input) over this range of about 50%. The highest coupling efficiency (laser output/electrical input) is obtained using a krypton gas filling with a lamp efficiency of about 40%. High-power pulsed discharge lamps present severe design problems due to the high mechanical stress on the lamp during operation. Continuous discharge lamps require intensive water cooling. The power supply uses capacitors to store the pulse energy using delay lines to shape the pulse wave form.

7.6
Ion Lasers

Ion lasers use ion transitions rather than excitation transitions, such as the helium–neon laser, and require considerably more energy to excite the ion transition, which is generally >10 eV. Several different ion lasers are available commercially, including argon, krypton, helium, cadmium and copper vapour. The higher energy of the ionization transition enables outputs of shorter wavelengths to be obtained.

The argon ion laser relies on emission from excited ions developed by a low-pressure arc discharge. The process is a two-stage process followed by ionization or excitation:

$$Ar \rightarrow Ar^+ + e$$
$$Ar^+ \rightarrow (Ar^+)^* + e$$

The construction of the laser (Figure 7.17) is similar to that of the helium–neon laser but operates at a higher partial pressure of helium of around 800 Pa (6 Torr) with output powers from 5 to 50 mW at 325 and 441 nm and efficiencies of about 0.05%. At low power output (<100 mW), a convection-cooled fused quartz tube is adequate to contain the discharge but at higher powers, ceramic or tungsten disks

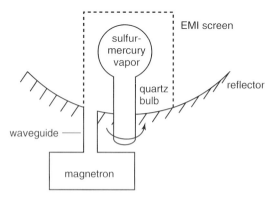

Figure 7.16 High intensity sulphur electrodeless lamp.

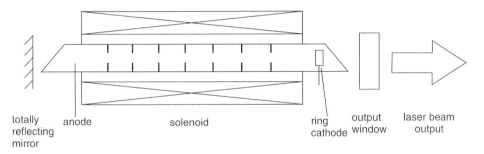

Figure 7.17 Construction of ion lasers.

with insulated surfaces are used to contain the discharge inside a ceramic tube externally cooled with water. The laser tube is only about 1 mm in diameter so as to stabilize the discharge on the optic axis and the ion flow along the bore towards the cathode results in a pressure gradient along the tube which is relieved by boring holes in the spacers or discs.

An axial magnetic field of about 0.1 T is used to obtain sufficient convergence of the plasma along the bore of the tube but is reduced to about 0.01 T at the cathode so that the current density at the arc root does not cause excessive sputtering and damage the cathode. The magnetic field causes spiral motion of conducting particles with an off-axis (radial) component of velocity. The magnetic field can be produced with permanent magnets or with a solenoid coil. Brewster windows allow external tuning of the laser so that the mirror coatings are not damaged by overheating.

The ion excitation process is inherently inefficient and requires current densities of the order of 10 A mm^{-2} at operating voltages of about 35–100 V DC. As a result, the laser requires cooling and at high powers (>3 W) water cooling is necessary. The output is up to about 30 W over the range 458–575 nm at a quantum efficiency of 0.05%. The process has a low efficiency, requiring current densities of the order of 2.5 A mm^{-2} at up to about 220 V and 30 A to achieve the excitation energy of 35.5 eV required.

The argon is contained at a pressure of 26.6 Pa (0.2 Torr) in a tube of about 3 mm diameter and 0.6 m long.

The helium–cadmium laser uses ionization transitions from cadmium vapour. Positively charged ions of cadmium vapour migrate towards the cathode (*cataphoresis*) but are condensed in a cold trap before they reach it. The Penning reaction between the metastable helium atoms and atomic cadmium is

$$He^* + Cd \rightarrow (Cd^*)^* + He + e \pm hf$$

7.7
Arc Interrupters

Arc interrupters have different operating requirements for different applications. A circuit breaker is required to interrupt a circuit at very high currents and volt amperes under full short-circuit conditions to prevent damage, but will normally only be used infrequently (if at all). Contactors are required to interrupt the load current in a circuit under normal load conditions but are used very frequently for different applications.

When a switch is opened, the energy stored in the inductance and capacitance of the circuit has to be dissipated in the resistance of the circuit and as a discharge across the switch contacts. At low voltages (<30 V), the switch design is relatively simple since the voltage available is insufficient to maintain the cathode and anode fall voltages necessary to sustain an arc discharge in air at atmospheric pressure, and the stored energy results in a transient back e.m.f. $e = -L di/dt$ and the energy in the circuit is dissipated in a spark at the contacts. At higher voltages, extinction of the discharge normally takes several half cycles (30–150 ms) due to the mechanical inertia of the moving contacts and the need to dissipate the stored energy in the circuit and to open the circuit at as low a current as possible. During this time, the current and stored energy in the circuit progressively decrease as the arc resistance increases and the arc is extinguished at a low current so that the back emf $e = -L di/dt$ is low when the voltage across the arc $\overline{V}_A > \overline{V}_S$ exceeds the voltage drop across the power source. The recovery time has to be sufficient to prevent the gap from breaking down again as the voltage across the contacts increases.

If the switch contacts are opened too slowly, a stable discharge may be formed which may cause damage to the switch, and in the case of a fault current may cause damage to other equipment. If the circuit is opened too quickly, the back emf will cause electrical breakdown elsewhere to dissipate the stored energy.

Various methods are available to increase the voltage between switch contacts such as by lengthening the arc, cooling the discharge using natural or forced convection in air or oil, use of electronegative gases such as sulphur hexafluoride, magnetic interaction of the self-magnetic field of the arc with its self magnetic field due to interaction with the current in the interrupter, or external magnetic fields (see Section. 3.2.1); vacuum arc circuit breakers break up the arc so that the arc voltage increases with the increased number of arc roots as well as increasing the discharge voltage at pressures below the Paschen minimum.

Compact circuit breakers operated in air with convection, arc chutes and magnetically driven arcs are adequate both as circuit breakers and as contactors. Oil-filled circuit breakers rely primarily on dissipation of the stored energy in the external circuit and in the discharge column to extinguish the arc. Rapid cooling of the arc aided by convection coupled with the excellent insulating properties of oil assist the extinction of the arc but have largely been superseded by air blast, vacuum or SF_6 circuit breakers.

Gas blast circuit breakers use forced convection from a jet of compressed air or an electronegative gas with a higher specific enthalpy to cool the arc and also remove vapour and ionized gas from between the switch contacts (Figure 7.18).

Circuit breakers using SF_6, which is electronegative and attracts free electrons and has a high electric strength, use a high-velocity flow of compressed gas to cool convectively the arc formed between hollow contacts as they separate. The arc may also be elongated by a magnetic field. The arc is drawn on the axis with the

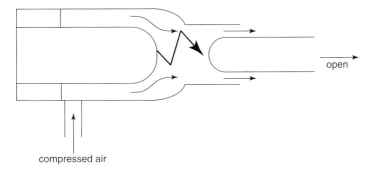

Figure 7.18 Air blast circuit breaker.

arc roots on the inside of the contacts and the gas is compressed by the opening contacts. SF_6 circuit breakers are established as the standard circuit breaker in AC transmission systems with voltages above 100 kV and are capable of operating at short-circuit currents of 80 kA and 800 kV.

7.7.1
Vacuum Circuit Breakers and Contactors

High-voltage vacuum switchgear is used to interrupt voltages up to 72 kV at fault currents as high as 63 kA and for contactors at voltages down to 11 kV. The Paschen curve in Figure 7.19 shows the breakdown voltage of dry air decreasing with gas pressure and separation (*pd*), from 3 kV mm^{-1} at atmospheric pressure down to a few tens of volts at 133 Pa (1 Torr). Initially the breakdown voltage decreases as the pressure decreases, but as the vacuum decreases below the Paschen minimum the breakdown voltage increases as the probability of cumulative collision decreases,

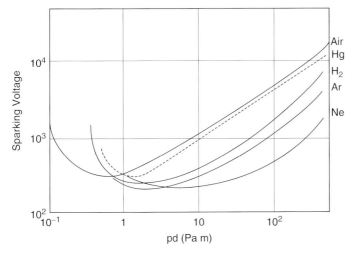

Figure 7.19 Paschen curve.

despite the higher particle energy. The high dielectric strength and the low thermal inertia of an arc at low pressure below the Paschen minimum has resulted in vacuum arcs being used for circuit breakers and in low-cost vacuum contactors for switching lower voltages and currents. The principle of the construction of the vacuum arc circuit breaker is shown in Figure 7.20.

The contacts are enclosed in a vacuum-tight enclosure at an operating pressure of 10^{-3}–10^{-5} Pa (7.52×10^{-6} – 7.52×10^{-8} Torr), well below the pressure corresponding to the Paschen minimum at 10^{-1} Pa (75.2×10^{-3} – 7.52×10^{-3} Torr). Both contact designs use the interaction of the self magnetic field of the current in the contacts to drive the arc radially otward and split it into lower cuurent spirals which are more easily extinguished.

One contact is connected to metal bellows so that it can be moved axially (Figure 7.20). The contacts are required to carry the full fault current for up to half

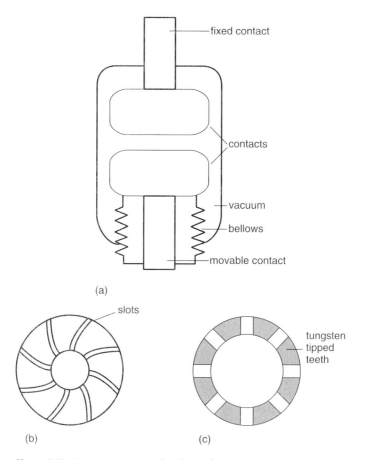

Figure 7.20 Vacuum arc circuit breaker and contacts; (a) Interrupter; (b) spiral petal contacts; (c) cup or concrete contacts.

a cycle. The contacts are opened close to zero current, minimizing the back emf ionized gas and vapour from the contacts in the gap at low pressure enabling the gap to recover to a high dielectric strength.

Vaporization of the cathode produces positive ions, which reduce the space charge and form the arc column, since there is insufficient gas to ionize and maintain an arc. The total arc voltage is about 15–20 V, mainly due to the cathode fall voltage with a column voltage gradient of the order of 0.1 V mm^{-1}. Estimated values of electron temperature are of the order of 4 eV at the cathode, with similar values in the column with electron densities of the order of 10^{20} m^{-3}.

7.8
Magnetoplasmadynamic Power Generation

Magnetoplasmadynamics (MPDs) is the interaction between an electrically conducting fluid such as a plasma with a magnetic field and can be used to generate power directly from the heat from hot gases. The MPD generator consists of a duct in which the plasma flows at about the speed of sound through a magnetic field. The Lorentz force $qu \times \overline{B}$ causes a flow of current and a voltage $e = \overline{B}u$ between the electrodes. To achieve the high level of electrical conductivity needed for the high current required, the hot gas is seeded with about 1% of potassium carbonate, generating an emf. For the generation of useful amounts of electrical power, very high magnetic flux densities of the order of 5 T are required. Although relatively little work has been carried out recently in the West on MPD generation, in Russia a complete generating plant has been installed near Moscow with a rated power of 20.5 MW, with a second installation of 500 MW planned.

7.9
Generation of Electricity by Nuclear Fusion

Nuclear fusion reaction requires the Coulomb repulsive forces between deuterium, which has nuclei with one proton and one neutron, and tritium, which has one proton and two neutrons, to be overcome for a sufficient time for them to be fused together and release energy.

The plasma is produced in a mixture of deuterium and tritium inside a toroid of up to 6 m external diameter at a pressure of <1.37 kPa (10^{-2} Torr), heated by a current of about 5 MA in the plasma which forms a single turn secondary (Figure 7.21). The power density is increased by coupling RF power and by the injection of high-power beams of neutral particles. The electron and ion energy achieved is about 15 keV (174 × 10^6 K) with an electron density of between 0.5 and 3.6 × 10^{19} m^{-3} and >10^{-1} degree of ionization and a number density $n_e = 10^{21}$.

The current results in a toroidal magnetic field. Ions and electrons follow the magnetic field lines. The magnetic field decreases across the toroid which, since the electrons and positive ions are travelling in opposite directions at different speeds,

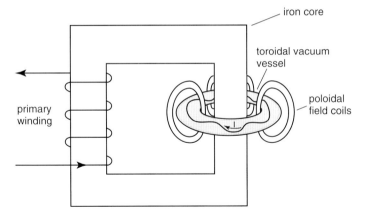

Figure 7.21 Schematic illustration of plasma toroid in a Tokamak fusion machine.

forces them to drift apart and an electric field develops, resulting in instabilities and in turn creating a variation in the magnetic field.

To overcome this instability, a magnetic field is generated which combines with the toroidal magnetic field due to the current to produce a helical field which stabilizes the plasma.

7.10
Natural Phenomena

Electric discharges in Nature are of general interest and are included here as examples of plasmas to illustrate some fundamental aspects of the behaviour of electric discharges.

7.10.1
Lightning

The Earth is surrounded by a region of ionized gas forming the Van Allen belt and the ionosphere. A positive potential of about 300 kV exists with respect to the Earth's surface. A predominantly positive ion density exists in the air near ground level caused by low-level emission of electrons (10^2 A m^{-2}) from the entire Earth's surface, which has a net value of the order of 700×10^6 positive ions per m^3. The effective resistance between the Earth and the ionosphere is about 200 Ω and the capacitance of the earth is about 18.2×10^{-9} F. The positive charge at the Earth's surface polarizes the cloud so that the bottom is negatively charged and the top becomes positively charged. Hot, moist air transfers more negative charge to the top of the cloud, increasing the voltage difference in the cloud, and to the Earth's surface. Most lightning strikes occur between clouds or within a cloud.

The propagation of a lightning strike occurs by leaders from the underside of the cloud or streamers from the Earth's surface until a conducting path is established.

7.10 Natural Phenomena

This is also accompanied by streamer-like discharges known as *sprites* reaching upwards from the top of the cloud. The air and positively charged ions are hottest closer to the Earth's surface and rise due to convection past the static negative ions to the top of the cloud. The effect of the temperature of the Earth's surface is critical, the number of thunderstorms increasing very rapidly with very small increases in the mean temperature. The intensity of charge increases with the depth of the cloud and deep nimbus clouds, each of which is separate, allow convective flow up the sides and may be 10 km deep. The negative ions at the base of the cloud above the Earth's surface remain under the influence of the excess positive charge. The process can be likened to the Van der Graaf high-voltage generator in which a very high voltage (MV) is obtained by building up a charge accumulated on separate water droplets.

Typically, the cloud ceiling is less than 10 km, hence the breakdown voltage at NTP assuming an electric stress of 3 kV mm^{-1} is 3×10^{10} V, whereas measured values are nearer 0.3–0.4 kV mm^{-1}, corresponding to 10^9 V. At 10 km, however, the pressure is only about 26.6×10^3 Pa (200 Torr) and the mean free path is about 13 times that at the Earth's surface, and the effect of water vapour also reduces the breakdown voltage.

The lightning strike occurs in a series of stages; however, the main strike normally occurs between Earth and the cloud over about 10^{-4} s with a current of about 10^4–10^5 A reached in tens of microseconds with temperatures as high as 30 000 K, with a channel diameter of 100 mm diameter, a rate of rise of current of the order of 10^{10} A s^{-1} and with an energy release of about 100 kJ m^{-1} or 50 kWh. The magnetic field H strength 1 m from the lightning strike is 10^4 A turns m^{-1} and the corresponding magnetic flux density $B = 10^4 \times 4\pi \times 10^{-7}$ T. The magnetic flux density decreases linearly with distance so that at a distance of 1 km it is 10^{-5} T.

The likelihood of a direct lightning strike can be reduced by developing an equipotential surface with an earthed metal surface or lightning conductor. The local charge and electric stress is greatly reduced and a corona discharge of up to 10^{-2} A flows over a region 50–100 m from the surface corresponding to a positive charge density of about 10^6 C m^{-3}, which reduces the local field gradient to less than the breakdown field in air (3 kV mm^{-1}). This prevents or reduces the effect of the development of a positive leader. The development of the positive space charge must take place slowly over several seconds to prevent a lightning strike, hence it cannot be used to protect aircraft travelling at several hundred miles per hour.

Ball-lightning, until recently often regarded as hysterical observation, can be explained in terms of known behaviour of electric discharges. Eye witnesses have reported slow moving palely glowing balls around 1 m in diameter at ground level lasting for several seconds before disrupting with explosive violence.

A number of different explanations have been advanced for ball-lightning and thunderbolts, for which there is sufficient evidence to indicate that they are not purely folklore. A plausible explanation for ball-lightning and thunderbolts can be put forward as follows. An ICP (RF or transformer coupled) can exist without electrodes. In the case of lightning, kinks occur presumably due to interaction of off-axis current with its self-magnetic field. A loop discharge can be formed caused by its jagged, multiple path, resulting in shunting of the column.

A lightning strike results in a rapidly changing electric field $dE/dt \propto 1/r^3$ and the moving current pulse in the leader stroke causes a magnetic pulse $di/dt \propto 1/r^2$, which causes voltage surges on overhead power lines (Chapter 6) and electromagnetic radiation $\propto 1/r$, resulting in electromagnetic interference within a frequency range from very low to very high frequency over a considerable distance. The rapidly changing magnetic field induces voltages of the order of 100 kV on overhead power lines and is the cause of many low-voltage (11 kV) distribution line faults even although a direct strike does not occur. The electromagnetic pulse is similar to that caused by a thermonuclear explosion which induces a high electromagnetic pulse radiating from it in a similar way but at a higher intensity than that from a lightning pulse. This is sufficient to destroy, by airborne interference, semiconductor devices many miles from the nuclear explosion.

Further Reading

Arc Welding, Spraying and Cutting

Guile, A.E. (1986) The electric arc, in *The Physics of Welding*, 2nd edn (ed. J.F. Lancaster), Pergamon Press, Oxford, pp. 120–148.

Hinkel, J.E. and Forsthoefel, F.W. (1976) High current density submerged arc welding with twin electrodes. *Welding Journal*, 175–180.

Houldcroft, P. and John, R. (1988) *Welding and Cutting: a Guide to Fusion Welding and Associated Cutting Processes*, Woodhead Publishing, Abington, Cambridge.

Kambara, M. Huang, H. and Yoshida, T. (2007) Recent progress in plasma spraying processing, in *Advanced Plasma Technology* (eds R. D'Agostino, P. Favia, Y. Kawai and H. Ikegami), Wiley-VCH Verlag GmbH, Weinheim, pp. 401–497.

Lancaster, J.F. (1986) The electric arc in welding, in *The Physics of Welding*, 2nd edn (ed. J.F. Lancaster), Pergamon Press, Oxford, pp. 228–305.

Electric Arc Melting

Barcza, N.A. (1986) The development of large scale thermal-plasma systems. *Journal of the South African Institute of Mining and Metallurgy*, **86** (2), 164–176.

Robiette, A.G.E. (1972) *Electric Melting Processes*, Griffiths, London.

Sayce, I.G. (1972) Plasma Processes in Extractive Metallurgy, Proc. Internat. Symposium on *Advances in Extractive Metallurgy and Refining*, Institution of Mining and Metallurgy, London, pp. 242–299.

Arc Gas Heaters

Barbier, M., Harlow, M., Oullette, R. and Pikul, R. (1978) Acetylene production by electric means, in *Electrotechnology, 2, Applications in Manufacturing* (eds R. Ouellette, F. Ellerbusch and P.N. Cheremisinoff), Ann Arbor Science Publishers, Ann Arbor, MI. pp. 293–333.

High-pressure Discharge Lamp and Lasers

Geens, B. and Wyner, E. (1993) Progress in high pressure sodium lamp technology. *Proceedings of the IEE, Part A*, **140** (6), 450–464.

Gunther, K. (2001) High-pressure plasma light sources, in *Low Temperature Plasma Physics* (eds R. Hippler, S. Pfau, M. Schmidt and K.H. Schoenbach), Wiley-VCH Verlag GmbH, Weinheim, pp. 407–431.

Sugiura, M. (1993) Arc discharge lamps: review of metal-halide discharge-lamp development 1980–1992. *Proceedings of the IEE, Part A*, **140**, 443–448.

Arc Interrupters

Attia, E.A. (1973) Arc stability characteristics of vacuum switches. *Proceedings of the IEEE*, **61** (8), 1156–1158.

Gerhard, C.J.O. (1976) High voltage switchgear. *Proceedings of the IEE*, **123** (I0R), 1053–1080.

Malkin, P. (1989) The vacuum arc and vacuum interruption. *Journal of Physics D: Applied Physics*, **22**, 1005–1019.

Reece, M.P. (1975) Physics of circuit-breaker arcs, in *Power Circuit Breaker Theory and Design*, IEE, Monograph Series, 17 (ed. C.H. Flurscheim), Peter Peregrinus, Stevenage.

Space Propulsion

Inutake, M., Ando, A., Tobari, H., and Hattori, K., *et al.* (2007) Development of physics issues of an advanced space propulsion, in *Advanced Plasma Technology* (eds R. D'Agostino, P. Favia, Y. Kawai and H. Ikegami), Wiley-VCH Verlag GmbH, Weinheim, pp. 435–448.

8
Diagnostic Methods

8.1
Introduction

Diagnostics is critical to the understanding of plasmas and the successful application of many plasma processes, particularly at low pressures, such as those used in the manufacture of computer chips and similar electronic components. This chapter deals mainly with diagnostics applied to the plasma rather than processes. Diagnostics applied to a process is highly specific to the application and is covered extensively elsewhere.

The principal parameters of interest are the number densities of neutral, charged and excited particles, the temperature and hence energy levels of electrons and ions, energy distributions of electrons and ions, sheath thicknesses and the distribution of electric and magnetic fields. Direct measurement using probes affects the plasma and the very wide range of operation of plasmas also means that few universally applicable methods are available. Non-contact methods, such as measurements of absorption or emission, do not significantly affect the plasma but are often complex and difficult to interpret.

8.2
Neutral Particle Density Measurement

The measurement of neutral particle densities is one of the simplest plasma measurements. Pressure is easily and accurately measured from around atmospheric pressure down to 0.1 mPa (0.752×10^{-6} Torr) with the McLeod gauge, in which a preset volume of gas is compressed and the resulting pressure is measured with a mercury column; however, this can result in contamination from the mercury. The Pirani gauge measures variations of heat transfer of a gas at low pressures using a resistance-heated filament or thermocouple from 10 Pa (75.2×10^{-3} Torr) down to 10^{-1} Pa (0.752×10^{-3} Torr), hot and cold ionization gauges down to about 10^{-8} Pa (75.2×10^{-11} Torr) and the cold cathode Penning gauge from 10^{-1} Pa (0.752×10^{-3} Torr) to 10^{-10} Pa (75.2×10^{-12} Torr).

Introduction to Plasma Technology: Science, Engineering and Applications. John Harry
Copyright © 2010 WILEY-VCH Verlag GmbH & Co. KGaA, Weinheim
ISBN: 978-3-527-32763-8

Neutral particle densities scale with pressure and temperature. Gas temperatures can be measured easily with a thermocouple, although this can disrupt a plasma. Indirect measurement of the neutral gas temperature can be made by measuring the vessel wall temperature, from the refractive index μ or sound velocity. When the density of neutral particles is low, such as in highly ionized plasmas, measurement of the attenuation of spectral lines is used.

8.3
Probes and Sensors

Probes are well-established diagnostic aids in most areas of manufacture but the sensitivity of plasmas to disruption of charge and hence charge equilibrium limits their use.

8.3.1
The Langmuir Probe

When a probe is inserted in a plasma, the surface of an electrically conducting probe superimposes an equipotential in the plasma region and a space charge is formed and flow of current may occur; an insulating probe builds up a surface charge which repels like charge.

A space charge is set up around it to preserve charge equilibrium in the plasma. The electrons have a much higher velocity than the ions and therefore hit the probe many more times and form a negative space charge of electrons due to the few slow-moving positive ions in the region, but unlike an electrode an isolated space charge does not carry a current. If the probe is a good conductor the sheath will be of positive ions, if the probe is an insulator the negative charge on the probe will result in a positive space charge in the plasma in front of the probe.

A simple Langmuir probe consists of a wire protruding from a glass sleeve (Figure 8.1) with its circuit. The Langmuir probe is used for measurements of electron temperature T_e, electron density n_e and the Maxwell distribution function

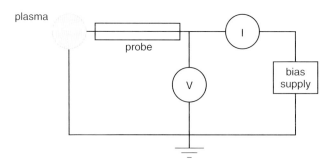

Figure 8.1 The Langmuir probe.

$f(u)$ in plasmas. The Langmuir probe partly overcomes the problem of disrupting the plasma by applying a bias voltage to the probe which nullifies the current flow from the plasma caused by collisions with electrons, so that a floating potential V_f is produced equal to the local potential in the undisturbed plasma. The use of Langmuir probes is usually limited to cold plasmas at pressures below about 10 Pa (7.52×10^{-3}), at which the mean free path is greater than the probe size, and the probe size is greater than the Debye length.

The probe is biased by a series voltage. A probe protected by an insulating sheath or a conducting sheath may become insulated by products of the process and can be biased with an AC bias voltage to overcome this. The effect on the local electric field in the plasma is minimized by making the probe diameter d very much less than the principal dimensions of the plasma and the probe face, $d \gg \lambda_D$.

The probe voltage is measured with respect to a reference potential, often the anode or earth. When a suitable electrode is not available, a second probe can be used. The current is limited by the external circuit impedance, which should be several orders of magnitude greater than the plasma, but low enough to prevent a charge developing on the surface. The electron current can be derived from Boltzmann's equation for the condition for equal electron and ion currents:

$$n_{e0} = n_e e^{-\frac{eV}{kT_e}} \tag{8.1}$$

where n_{e0} is the electron density at the probe surface and n_e is the electron density at the plasma boundary of the sheath.

The characteristic I–V curve for a Langmuir probe with the current on the vertical axis and the voltage on the horizontal axis is shown in Figure 8.2. Over the electron and ion saturation regions AB and DE the probe acts in a similar way to an electrode with negative (cathode) or positive (anode) bias.

At B, the positive space charge at the probe surface is equal to the floating potential at which no current is taken from the bias supply. The measured value is below the ambient potential since more electrons arrive at the probe than ions because of the greater velocity of electrons. Between C and D at the knee of the curve, the positive space discharge decreases until the space charge is reduced to zero.

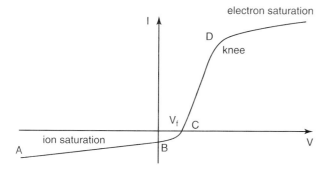

Figure 8.2 The Langmuir probe characteristic.

If the characteristic curve is drawn with the current on a logarithmic scale and the voltage on a linear scale, then

$$\ln i_e = \ln i_{es} - \frac{eV}{kT_e} \tag{8.2}$$

where i_e is the electron current at a bias voltage V negative to the plasma and i_{es} is the random electron flow across the probe face at saturation when $V = 0$.

The linear part of the curve C–D has a slope e/kT_e, from which the electron temperature can be found. The electron density can be found from the equation for the random electron current:

$$j_{er} = n_e e \left(\frac{kT_e}{2\pi m_e} \right)^{\frac{1}{2}} \tag{8.3}$$

putting i_{es}/A for j_{er}. The plasma potential is determined from the voltage at the knee of the curve where there is no space charge. The point D is not easily determined but can be more accurately defined by electronic differentiation of the signal.

8.3.2
Magnetic Probes

Magnetic probes can be used to measure current, current density or motion of a discharge. One of the simplest and universally available probe methods is the magnetic probe, similar to magnetic probes used for current measurement (Figure 8.3).

The magnetic probe measures the voltage due to a change in magnetic flux $V = -NA(d\phi/dt)$. The simplest magnetic probe is an air-cored coil which surrounds or is immersed in the plasma and the output is measured with an oscilloscope or the root mean square (r.m.s.) or mean value of the output can be integrated electronically. The output voltage can be increased by using a ferrite core.

If the coil has an area A, the voltage can be written in terms of the flux density $V = -NA(dB/dt)$.

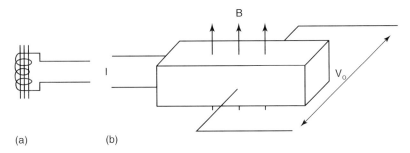

Figure 8.3 Magnetic probes: (a) magnetic coil probe wound on a ferrite core; (b) Hall probe.

Coil probes are limited to fluctuating magnetic fields; Hall effect probes (Figure 8.3b) can also measure steady magnetic fields. Their frequency response is limited, however, and the probes are sensitive to electrical noise and are less robust than coil probes.

8.4
Optical Spectroscopy

Optical methods of diagnostics such as emission spectroscopy, absorption spectroscopy and laser-induced fluorescence (LIF) give useful information. The intensity of a spectral line is a function of the temperature of the emitted species and the wavelength corresponds to the atom or molecule. The relative intensity method is in theory one of the simpler spectroscopic methods, but it is subject to errors if the discharge is not homogeneous or optically thin. Absolute intensity measurements can be made but are more complex.

Spectra observed in a plasma are listed in Table 8.1.

8.4.1
Optical Emission Spectroscopy

Planck's equation for the intensity of spectral emission from a source for unit solid angle is

$$I'(\lambda, T) \alpha \frac{2hc^2}{\lambda^5} \left(\frac{1}{e^{\frac{hc}{\lambda kT}} - 1} \right) \tag{8.4}$$

The frequency and wavelength of emission are proportional to the difference in energy between the upper and lower states such that

$$hf = \zeta_2 - \zeta_1 \tag{8.5}$$

where E is the energy (eV), h is Planck's constant (6.626×10^{-34} J s), f the frequency (Hz) and λ the wavelength (m).

Table 8.1 Spectra observed in a plasma.

Particle	Degree of freedom	Type of spectrum	Spectral region
Atoms or ions	Electronic excitation	Line	UV–visible–IR
	Ionization	Continuum	UV–visible–IR
	Translation	Line profiles	–
Electrons	Recombination	Continuum	UV–visible
	Free–free transitions	Continuum	IR
Molecules	Rotation	Line	Far-infrared
	Vibration–rotation	Band	IR
	Electronic excitation	Band systems	UV–visible–IR

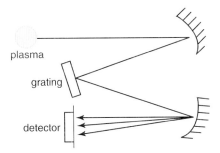

Figure 8.4 Optical spectrometer.

An optical spectrometer is illustrated schematically in Figure 8.4. Light is collected from the plasma using a lens or fibre optic and divided into its spectrum by an interference grating or by optical filters. A photodiode, a charge-coupled detector or a photomultiplier is used for the detector depending on the spectral range and sensitivity required. The relative line strengths of spectral lines can be obtained by comparison with a reference emission source for the same conditions; any differences between the source and reference lead to inaccuracies.

Emission from atoms from electronic transitions to the ground state are often well defined and can be used to identify neutral particles and ions of light-emitting species from tables of emitted wavelengths. The spectra of molecules and ions are more complex than for atomic transitions from the ground state. In principle, the intensity of the line spectra should enable the concentration of an excited species to be determined. However, if any changes are made that affect the excitation efficiency or the plasma parameters, this is no longer true.

Actinometry involves the addition of a known amount of an impurity with known spectra to produce suitable spectral lines of approximately the same energy to be analysed. Two spectra can be compared under the same conditions and enable the number density of the plasma spectra to be compared. A gas can be added to discharges to produce suitable spectral lines. The gas should have a high ionization potential so that the temperature of the discharge is not lowered appreciably by its addition (the reverse of seeding). If changes are made to the plasma, and if the inert gas is not affected the parameters of the plasma can be deduced from the added gas.

Molecules have correspondingly more electronic states and vibrational and rotational states so that the emission overlaps in bands that broaden as the pressure increases. In addition to line emission, chemiluminescence, ion–neutral collisions, fluorescence, phosphorescence, relaxing metastable states and other processes may also occur.

Where line broadening occurs in hot, highly ionized plasmas, the velocity of the emitted ions or atoms can be determined from Doppler broadening and the density from Stark or pressure broadening (high-density collisions broaden the frequency).

High temporal and spatial resolutions of neutral and ion particle densities have been achieved by LIF. A tuneable laser, such as a dye laser, is used to excite metal vapour in the discharge so as to fill an upper fluorescent state. The concentration of particles is proportional to the intensity of the emission from the upper level and can be determined if the transition probabilities and statistical weighting of the energy levels are known.

Spectroscopic observations measure a value of intensity which is integrated over the thickness of the discharge rather than radiation from a point source. Because of the steep temperature gradients associated with high-pressure discharges, the radiation detected in any spectroscopic diagnostic technique is the integral of that emitted by various elements of the discharge at greatly differing temperature. If the geometry of the discharge is uncertain, this can cause errors in the measurement of the radial intensity profiles of free-burning arcs and arcs from plasma torches, which are assumed to be cylindrically symmetrical, because of the different thicknesses of the discharge observed as the cross-section is scanned from the side. The true radial intensity or temperature distribution can be resolved from side-on observations of an S± emission line by the Abel inversion.

8.4.2
Absorption Spectroscopy

Absorption spectroscopy is a relatively simple and direct technique for the detection of atomic transitions from the ground state and can be quantitative, providing information on intensity, linewidths, wavelengths and spectral lineshape, but has relatively poor resolution. It is limited in species to those that absorb emission from laser or other light sources. Sensitivity is increased using frequency-modulated laser absorption spectroscopy or intracavity laser absorption spectroscopy.

If the plasma is only weakly ionized and above the collision frequency, $\omega_s > \omega_c$, the wave will be propagated and may be absorbed if the electrical conductivity is sufficient. Absorption of the energy occurs exponentially with distance from the surface and if the conductivity is too low propagation through the plasma occurs whereas at high electron number densities it will be propagated.

The attenuation by the plasma is exponential:

$$I = I_0 e^{-\alpha t} \tag{8.6}$$

where α is the attenuation coefficient.

Absorption methods enable the absolute number densities of the absorbing species to be determined. Since the detection methods for absorption do not, in general, depend on the transport or radiative properties of the upper state, species of interest which do not fluoresce can still be detected.

8.4.3
Scattering Measurements

Measurements of scattering of radiation in cold plasmas can be used to measure electron temperature and density and by using well-collimated laser beams, temperature profiles can be measured. Scattering of the incident light results from oscillation of the electrons caused by the incident light, which causes them to radiate in all directions. The electron density is determined from the difference in the wavenumber k for $\lambda_D \gg 1$, that is, the difference in wavelength is much less than the Debye length, and then the shielding effect of electrons around ions can be neglected and the scattering measured is from the individual electrons. If an electron has a velocity u, the radiation scattered from it is Doppler shifted in frequency by an amount $k\bar{u}$ so that the spectral width of the scattered radiation is related to the electron temperature. If $k\lambda_D \geqslant 1$, the difference in wavelength is equal to or greater than the Debye length and the scattered radiation corresponds to that from the shield of electrons shielding positive ions and the spectral width of the scattered radiation depends on the velocity distribution of the ions. There are also fluctuations at frequencies corresponding to the plasma frequency, producing peaks in the scattered radiation. The scattering is referred to as *collective scattering* since it is dependent on the collective fluctuation of electron density rather than that of individual electrons.

8.5
Interferometry

Disturbances in gases such as caused by pressure differences and temperature gradients result in interference patterns caused by changes in the refractive index of the medium. Interferometry is used to measure pathlength by counting the interference fringes at half-wavelength intervals $n\lambda/2$. The principle is illustrated schematically in Figure 8.5. The interferometer is set up with the two mirrors equi-spaced so that the pathlengths are the same. If a change is made to the optical pathlength in one branch, for example if the refractive index over one of the paths changes, the reflected wave is delayed. The frequency of visible light is many orders of magnitude greater than the plasma frequency so that phase shift measurements can be made with conventional optical interferometers (Figure 8.5).

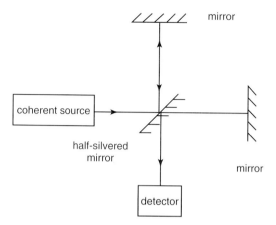

Figure 8.5 Interferometer.

Interference patterns can be used to provide data on pressure gradients in a plasma. Changes in the refractive index within glow and arc discharges can be shown using continuous or pulsed coherent lasers with visible outputs using Schlieren photography. An approximately parallel beam of light is passed through the discharge and compared with an undisturbed reference beam from a coherent light source. The diffracted rays are separated from the illuminating beam by a knife edge at the focus of the mirror and focused on to a screen showing the relative variation in refractive index as dark and light regions, enabling spatial measurements to be made. Although Schlieren methods are possible without a laser, the monochromatic and coherent output of a laser simplifies its use and pulsed lasers have been used to investigate transient discharges.

8.5.1
Microwave Interferometer

For waves to propagate through a plasma, the supply frequency must be greater than the collision frequency, $\omega_s > \omega_c$, and less than the plasma frequency, $\omega_s < \omega_p$. The phase velocity in the plasma is

$$\frac{\omega}{k} = \frac{c}{\left(1 - \omega_p^2/\omega^2\right)^{\frac{1}{2}}} \tag{8.7}$$

This is faster than the velocity of light but the group velocity is less. The microwave signal has a longer wavelength in the plasma than in air and the phase changes with increase in the plasma density. Figure 8.6 illustrates the principle of a microwave interferometer. Like the optical interferometer, the microwave interferometer compares the signal that passes through the plasma with the signal from the plasma source and measures the difference in wavelength by adjusting the attenuation and phase between them. Alternatively, the signal is reflected back through the plasma so that the beam experiences twice the phase shift, which increases its sensitivity and eliminates any change in phase shift of the reference signal due to changes in room temperature, which might affect the pathlength.

The interferometer can be calibrated by increasing the plasma intensity from zero since a null point every half wavelength occurs. If $\omega \approx \omega_p$, the phase shift is

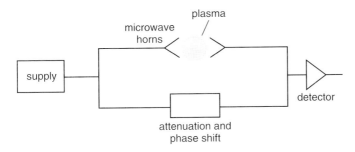

Figure 8.6 Microwave interferometer.

approximately linear and proportional to the number n of null points corresponding to $\lambda/2$. The plasma reduces the pathlength of the wave so that the path difference is determined from $\mu < 1$ and

$$\mu \approx \sqrt{1 - (\omega_p/\omega)^2} \tag{8.8}$$

If the signal of an electromagnetic wave is propagated through the plasma at the resonant frequency, it can be used to determine the number density n_c since the critical density

$$\omega^2 = n_c e^2/\varepsilon_0 m \tag{8.9}$$

If $k_0 = \omega/c$, the phase shift is $\Delta\phi = \int (k_0 - k_1) dx$, where k is

$$k = k_0 \left(1 - \frac{n}{n_c}\right)^{\frac{1}{2}} \tag{8.10}$$

where $\omega_p^2/\omega^2 = n/n_c$, and the phase shift is

$$\Delta\phi = k_0 \int \left\{ 1 - \left[1 - \frac{n(x)}{n_c}\right]^{\frac{1}{2}} \right\} dx \tag{8.11}$$

A limitation of many spectroscopic methods is that the measured value is integrated over the depth of the plasma rather than the local density; however, this can be determined from the Abel inversion, which gives the parameter of a cylindrically symmetric plasma as a function of the radius in a plane transverse to the plasma

$$f(r) = -\frac{1}{\pi} \int_r^a \frac{dF}{dr} \frac{dr}{(y^2 - r^2)^{\frac{1}{2}}} \tag{8.12}$$

8.6
Mass Spectrometry

Mass spectrometry measures the presence of different species by separating ions by their mass to charge ratio, m/z ($z = ne$). A sample of the plasma or gas is extracted through a differentially pumped region, ionized and collimated. The beam of ions is accelerated into the mass spectrometer. One form of mass spectrometer uses a magnetic field to deflect a beam of ions (see Section 3.3) which if they are travelling at the same velocity are deflected in a circular path with a radius proportional to their mass when they enter a magnetic field perpendicular to their direction of

$$r = \frac{mu}{eB} \tag{8.13}$$

The quadrupole mass spectrometer uses four shaped electrodes (quadrupoles) with opposite electrodes of the same polarity. (Four electrodes are needed to determine positive and negative ions separately.) Positive ions entering on the common axis (Figure 8.7) are attracted and drift to the negative electrodes and negative ions drift to the positive electrodes. An RF voltage is superimposed on the DC supply

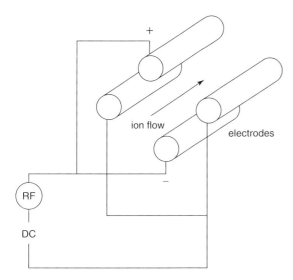

Figure 8.7 Quadrupole mass spectrometer.

(which may be many times larger than the DC voltage) so that the voltage at the positive DC electrode is increased for part of the RF cycle only and vice versa for the negative electrode. The lighter ions are deflected by the RF field; however, the heavier ions are less affected. The variable-frequency signal gives a sensitive method of controlling the net deflection enabling differentiation between small differences in mass. The quadrupole mass spectrometer can also be used to detect neutral particles if they are preionised before entering the electrode region and a similar process can be used to detect the particle size of sub micron particles.

The effective deflection voltage can be written as $V_{DC} = kV_{RF}$ and the mass resolution as $\Delta M = k_1 \xi f^2 l^2$, where k_1 is a constant value of k, f the RF frequency, l the length of the quadrupole and ξ the ion energy at the entrance to the quadrupole. By suitable selection of the axial velocity of the ions and the relative values of the DC and RF voltages, both positive and negative ions can be differentiated in terms of mass and charge and hence the molecule determined.

8.7
Electrical Measurements

Measurement of the parameters of electric discharges and plasmas presents a number of problems not normally encountered. Spatially stable discharges such as discharge lamps which have both temporal and spatial stabilization using power frequency AC have regular waveforms but are non-sinusoidal. At high frequencies, conventional digital meters are unreliable and the effect of inductance and capacitance may cause errors.

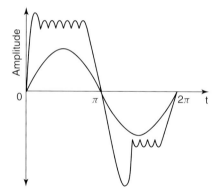

Figure 8.8 Non-repetitive AC waveform.

8.7.1
Electrical Instrumentation

The voltage and current of a plasma often fluctuate (Figure 8.8). The values of voltage and current of a fluctuating waveform are defined as follows. The average or mean value of voltage is

$$V_{av} = \frac{1}{T}\int_0^T v\,dt$$

For a sine wave, although over a full cycle the true average is zero, the average over one half cycle is given by $V_{av} = (2/\pi)\,\hat{v}$ and the r.m.s. voltage

$$V_{rms} = \sqrt{\frac{1}{T}\int_0^T v^2\,dt}$$

The corresponding values of current are defined in the same way.

The power $W = (1/T)\int_0^T vi\,dt$ and the energy $E = \int_0^T vi\,dt$ so that for a continuous sinusoidal waveform with an r.m.s. voltage V and r.m.s. current I the peak voltage $\hat{v} = \sqrt{2}V_{rms}$, peak current $\hat{i} = \sqrt{2}I_{rms}$, power $P = V_{rms}I_{rms}$ and energy $\zeta = V_{rms}I_{rms}t$.

The r.m.s. values are most commonly used since they relate to useful power from the plasma and are related to other measures such as the average value by geometric form factors, $V_{rms} = 1.11\,V_{av}$ for a sine wave.

Similar relations can be derived for other continuously fluctuating waveforms such as square and triangular waveforms.

Figure 8.9 shows the basic circuit used for measuring voltage and current. If the voltage and current are stable analogue or digital instruments calibrated for DC, average or r.m.s. waveforms can be used. Analogue and digital instruments calibrated for average or r.m.s. values can also be used with stable AC discharges if the waveform is repetitive from cycle to cycle, such as in a discharge lamp. However, calibration may be necessary for a non-sinusoidal waveform.

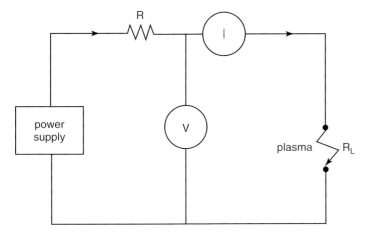

Figure 8.9 Voltage and current measurement.

Where the voltage and current vary from cycle to cycle in processes such as arc welding or arc melting where the arc can move and the electrode separation is not fixed. Although measurements of AC current are possible, the arc voltage may fluctuate over a wide range from near open circuit to short circuit, exacerbated by the effects of ignition and extinction peaks.

When the plasma voltage and current fluctuate with time, measurements made using digital or analogue instruments are normally useless and the only true measurement of basic quantities, such as V, I, W and frequency, can be made from waveforms obtained using an oscilloscope.

At high frequencies and high voltages, voltage and current measurements need careful interpretation due to alternative capacitive or corona paths, and this is particularly a cause of error where connection to the electrodes is not possible, such as in barrier discharges. At high frequencies, the capacitance with respect to earth may be a serious source of error. To avoid this, the current is measured close to earth potential by connecting the current transducer in the earth line as close to the electrodes as possible (Figure 8.10).

8.7.2
The Oscilloscope

Oscilloscopes are capable of measuring very high frequencies and rapid pulses. The transient response (rise time) of the oscilloscope (and probe if used) must be sufficient to analyse rapid changes in voltage and current, such as ignition and extinction events.

The *rise time* of a pulse is defined as the interval between 10 and 90% of the final value and for a sinusoidal waveform the frequency bandwidth is defined between 3 dB points at which the signal has decreased to $1/\sqrt{2}$ of its amplitude. The rise time is proportional to the time constant CR of a circuit. For a capacitive circuit the rise time $t_r \approx \tau \ln 9 \approx 2.2\, CR$.

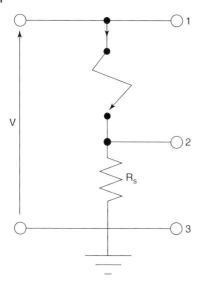

Figure 8.10 Current measurement at high voltage.

The bandwidth BW,

$$f_H = \frac{1}{2\pi CR}$$

$$t_r = \frac{1}{2\pi f_H} r \approx 0.35/BW$$

Hence for an oscilloscope of 300 MHz, the rise time is 1 ns.

Reflections can occur from a line which is long compared with its wavelength when a line is not correctly matched. If the reflected wave is out-of-phase with the transmitted wave, the voltages of the transmitted and reflected waves add or subtract and can damage the power supply or cause instrumentation errors if not correctly matched to the source or instrument. The impedance of a line is matched if no reflection from the end of the line occurs. Over distances $\ll \lambda$ a reflected wave is close to being in-phase with the transmitted wave but for general use a high-impedance wave so that a significant increase in the voltages does not occur.

The high voltages associated with discharges and plasmas often require attenuation and isolation. For general use, a high-impedance probe using a resistive divider with a small capacitor to compensate for the cable capacitance so as to give the correct pulse response is utilized. The capacitance of the probe can be varied using a square wave to see if there is overshoot or slow edge on the waveform displayed.

8.7.3
Electrical Measurements Using Probes

The high voltages associated with discharges and plasmas often require attenuation and isolation. Attenuation of a signal at high voltages results in high power

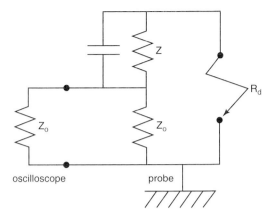

Figure 8.11 Oscilloscope and probe impedances.

dissipation and to avoid this capacitors are often used. Capacitor probes to attenuate the signal combined with resistors to balance charge are used to attenuate and match it to an oscilloscope. At very high voltages, adequate insulation safety, prevention of corona and voltage breakdown become important and oil-filled probes may be used. Use of a probe introduces errors in measurement due to the impedance of the probe (Figure 8.11). The coaxial cable has negligible mutual inductance and low resistance except when skin effect becomes significant and its impedance (typically 50 Ω) is approximately constant and equal to $\sqrt{L/C}$, independent of the length of the cable although a time delay may be introduced. The effect of the load on the oscilloscope, which typically has an input impedance of 1 MΩ, is negligible but may be significant, particularly if the impedance at the point of measurement Z_L is several times the probe impedance Z_P. The impedance of the probe may also result in an increase in circuit reactance and even a change in frequency, and in some cases introduction of a probe may even cause or stop resonance.

The simultaneous measurement of two or more quantities, such as voltage and current, is possible with an oscilloscope. However, the inputs are not usually isolated and share a common connection through the earthed chassis of the oscilloscope, which is normally connected to the mains supply earth (Figure 8.12). Where it is not possible to have a common connection, an isolated or a differential input is necessary.

If one side of the signal input to the oscilloscope is itself connected to earth, an earth loop around the mains supply exists, which will cause errors and results in an electric shock hazard. This is made even more dangerous if one side of the signal is connected to the neutral point of the supply. For example, if a variable autotransformer is used to supply the discharge circuit, the neutral is then earthed in addition to its main earth connection, usually at the substation, and a potentially dangerous situation exists.

At high voltages, where the impedance of a probe may be unacceptable or where safety is a problem, an optical isolator can be used to couple the output to the

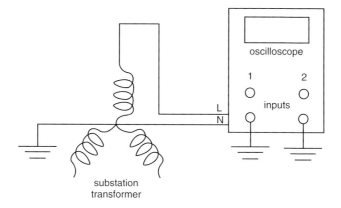

Figure 8.12 Oscilloscope earth return path to the substation transformer.

oscilloscope. This may use a transformer or an electro-optic transmitter and optic fibre and detector where the highest levels of isolation are needed for safety.

The measurement of phase shift between the plasma voltage and current is often a useful diagnostic tool.

8.7.4
Current Measurement

Several different current transducers can be used for current measurement (Figure 8.13). A simple and reliable method of measuring current is by measuring the voltage drop across a current shunt of known resistance, inductance and capacitance, which is connected in series with the supply (Figure 8.13a). Limitations of current shunts are the difficulty of manufacturing an accurate low resistance with low inductance for use at high current and the effects of stray inductance and capacitance at high frequencies. Resistive current shunts do not provide electrical isolation.

Current transformers have the advantage of dissipating low powers and providing electrical isolation, but the output of iron- or ferrite-cored transformers must be correctly matched and never used with an open circuit on the secondary. Iron-core current transformers (Figure 8.13b) can be used over the range 25 Hz–1 kHz in which the current carrying conductor is passed through an aperture in the transformer core to form the primary. The secondary current is given by the transformer relation:

$$I_S = \frac{N_P}{N_S} I_P$$

Current transformers are used for current measurements up to more than 10 kA at power frequencies and at lower currents using ferrite or air cores at frequencies up to 100 MHz.

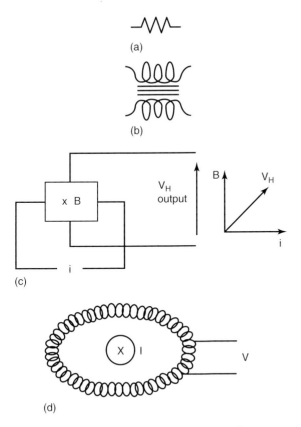

Figure 8.13 Current transducers. (a) resistance, (b) current transformer, (c) Hall current probe, (d) Rogowski coil.

The Rogowski coil current transformer has a solenoid winding bent into a circle so as to form a torus with the winding ending at the start. In this way, only the flux density within the solenoid is measured, which screens the signal from interference from outside the toroid. Only the magnetic flux in the centre of the torus is enclosed by the solenoid winding and minimizes pick-up of electrical noise. It can also be used for measurement of magnetic flux.

The conductor passes through the centre and, if a ferrite core or air core is used, the coil has a fast response and can be used up to 100 MHz.

Hall effect probes (Figure 8.13c) can be used for current measurements over the range DC to 100 MHz up to about 100 A. The rate of diffusion of charges in a Hall effect semiconductor is such that a magnetic field across a current-carrying probe deflects the current so that its pathlength and voltage drop change. The magnetic field can be produced by a transformer core surrounding the conductor with the probe mounted in the magnetic circuit. The output is isolated and normally requires amplification.

Further Reading

Overview

Hutchinson, H.I. (2002) *Principles of Plasma Diagnostic*, 2nd edn, Cambridge University Press, Cambridge.

Lochte-Holtgreven, W. (1995) *Plasma Diagnostics*, American Institute of Physics, New York.

Probes

Pfau, S. and Tichy, M. (2001) Langmuir probe diagnostics of low-temperature plasmas, in *Low Temperature Plasma Physics* (eds R. Hippler, S. Pfau, M. Schmidt and K.H. Schoenbach), Wiley-VCH Verlag GmbH, Berlin, pp. 131–172.

Swift, J.D. and Schwar, M.J.R. (1970) *Electrical Probes for Plasma Diagnostics*, Elsevier, New York.

Spectroscopy

Cabannes, F. and Chapelle, J. (1971) Spectroscopic plasma diagnostics, in *Reactions Under Plasma Conditions* (ed. M. Venugopalan), Wiley-Interscience, New York, pp. 367–469.

Ropcke, J., Davies, P.B., Kaning, M. and Lavrov, B.P. (2007) Diagnostics of non-equilibrium molecular plasmas using emission and absorption spectroscopy, in *Low Temperature Plasma Physics* (eds R. Hippler, S. Pfau, M. Schmidt and K.H. Schoenbach), Wiley-VCH Verlag GmbH, Weinheim, pp. 173–197.

Schmidt, M., Foester, K.H. and Basner, R. (2001) Mass spectrometric diagnostics, in *Low Temperature Plasma Physics* (eds R. Hippler, S. Pfau, M. Schmidt and K.H. Schoenbach), Wiley-VCH Verlag GmbH, Weinheim, pp. 199–227.

9
Matching, Resonance and Stability

9.1
Introduction

The operating parameters of a plasma are determined by the interaction between the plasma and the power supply. A plasma generally has a negative dynamic resistance, that is, the resistance decreases with increase in current and the current is limited only by the series resistance or impedance of the supply circuit. When the plasma reaches an equilibrium value it is said to be *stabilized*, also referred to as *controlled* or *regulated*. Obtaining the required operating conditions by stabilizing the plasma is essential for successful operation of the process.

Resistance can be used to stabilize a plasma but results in a waste of energy, with over half the total power being dissipated as heat, and reactive stabilization using inductors or capacitors is normally used. The negative dynamic resistance of the plasma and the stabilizing reactance together with stray inductance and capacitance of the circuit can cause resonance in the supply circuit as a result of fluctuations in the plasma.

Matching the plasma supply is important for obtaining the designed output from the supply: the load must be matched to the supply. Matching is also needed to avoid unwanted reflections of signals.

9.2
The Plasma Characteristic

A plasma has resistance but the dynamic resistance is normally negative or zero, even though the discharge characteristic may have a positive dynamic resistance over a limited range of current.

The generalized discharge characteristic is shown in Figure 9.1 with the current on a logarithmic scale so that the full range of the discharge can be shown. The current is shown by convention on the x-axis, although the voltage (or rather the electric field, E) is the independent variable.

Figure 9.2 shows a plasma supplied from a DC supply with an open-circuit voltage V_{oc} and short-circuit current I_{sc}. The curvature of the characteristic is shown

Introduction to Plasma Technology: Science, Engineering and Applications. John Harry
Copyright © 2010 WILEY-VCH Verlag GmbH & Co. KGaA, Weinheim
ISBN: 978-3-527-32763-8

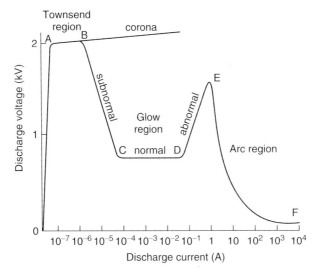

Figure 9.1 Generalized discharge characteristic (semi-logarithmic scale).

exaggerated to illustrate the effect. The series resistor R_s is used to limit the current in the plasma, which has a resistance r_d. An impedance Z or reactance X could be used to stabilize the plasma if it was supplied from AC; however, this makes the solution of the circuit equations more complicated.

The equilibrium discharge voltage and current must lie on the discharge characteristic at the point where the load line intersects it. The operating point at the intersection of the load line with the discharge V–I characteristic for a DC plasma is

$$V_{oc} = IR_s + f(i_d) = IR_s + V_d \tag{9.1}$$

The load line normally has two points of intersection A and B with the discharge characteristic (Figure 9.2). The steady-state resistance of the plasma at B is $R_d = V_d/I_d$. At point A on the plasma characteristic, for a decrease in current in the plasma $dv_d/di \geqslant R_s$ the voltage change $\delta v_d \leq R_s \delta i$, hence there is additional voltage available to increase the current further. This continues until point B is reached, when if the current tries to increase as $dv_d/di \leq R_s$ there will be a reduction in voltage, returning it to point B.

Hence for stability:

$$\frac{dv_d}{di} \leq R_s \tag{9.2}$$

is required, known as the *stability criterion*. The minimum voltage for a stable discharge is $2I_d R_s$ and $V_{oc} \geqslant 2V_d$.

An increase in the supply voltage V_{oc} shifts the load line vertically upwards and an increase in the series resistance R_s increases the slope. Changes in the discharge such as its pathlength, which change the voltage between the electrodes, also shift the discharge characteristic vertically.

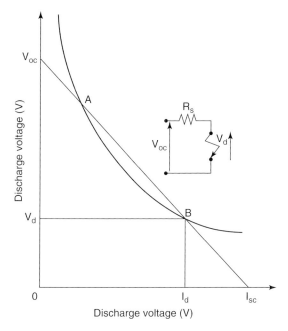

Figure 9.2 Interaction of the load line with the plasma characteristic.

The power in a discharge stabilized by a series resistor is

$$W = I^2 R_d = \frac{V_o^2 R_d}{(R_d + R_s)^2} \tag{9.3}$$

The condition for maximum power is obtained by differentiating the above equation and equating it to zero:

$$\frac{dW}{dR_d} = \frac{2V_o^2 (R_s - R_d)}{(R_s + R_d)^2} = 0 \tag{9.4}$$

For maximum power in the plasma, $R_d = R_s$, for which the current is a maximum for stable operation, the power in the plasma is half the maximum power available from the supply. For maximum use of the supply plasma, $R_d = R_s$, the voltage drop across the plasma $V_d = \frac{1}{2} V_s$ and the power in the plasma is $V_{oc} \times I_{sc}$.

At maximum power, the plasma characteristic is tangential to the load line and the plasma voltage V_d is half the supply voltage V_s. Any further increase in the plasma voltage results in extinction of the plasma. Operation at a higher ratio of plasma voltage to open-circuit supply is possible and improves stability.

The effect of the series stabilizing resistance R_s changes the source from a constant voltage source more closely to a constant current source where $I = V_{oc}/(R_s + R_d)$, with $R_s > R_d$ so that $I \approx V_{oc}/R_s$ and fairly large variations in discharge resistance result in only small changes in current.

Stabilization may also be used to limit plasma fluctuations supplied from DC power sources by connecting an inductor connected in series with the DC output.

Iron or ferrite cores are used at frequencies up to about 100 MHz; saturation of the core (see Section 10.2) is reduced by including an air gap in the transformer core. The effect of fluctuations in the plasma is reduced if the time constant L/R of the circuit τ_c is longer than the time constant of the plasma fluctuation τ_d.

9.3
Stabilizing Methods

In the past, batteries have been used to supply electric discharges and enable a ripple-free stable discharge to be obtained. DC generators have been used for high current arcs, are robust and allow overload for short periods, but are cumbersome. Figure 9.3 shows different methods of stabilizing plasmas. Figure 9.3a shows a series resistance used to stabilize a plasma. Resistors are needed for stabilizing DC and more than half of the power in the discharge is dissipated in the stabilizing resistor.

Simple and readily available methods for stabilizing AC and DC electric discharges include water-cooled resistors, incandescent light bulbs, metal-sheathed heating elements and low-inductance thick-film resistors. At high currents, electrodes immersed in water baths, expanded mesh metal elements and metal strip wound on edge can be used.

9.3.1
Reactive Stabilization

An AC plasma behaves in a similar way to a DC plasma at low frequencies (although if reactive stabilization is used the load line is no longer a straight line) and the circuit equations for a stable plasma are more complicated. AC electric discharges can be stabilized by reactance without the power loss due to resistive stabilization.

Figure 9.3b,c show a series inductor and a capacitor used for stabilizing a discharge. DC power supplies using rectified AC can be stabilized by connecting an inductor or capacitor in series with the input to the rectifier (Figure 9.3d).

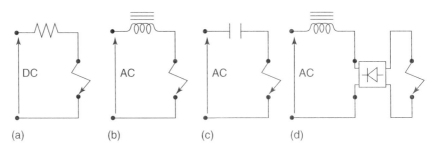

Figure 9.3 Simple stabilization circuits. (a) DC resistance; (b) AC inductor; (c) capacitor; (d) DC inductor.

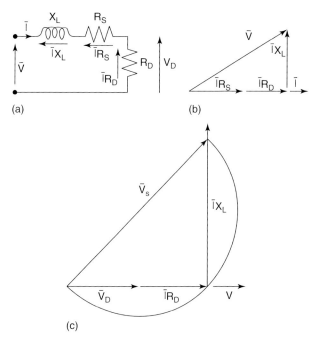

Figure 9.4 Vector diagram for a plasma stabilized by an inductor (a) inductive circuit, (b) vector diagram for inductive circuit, (c) circle diagram from (b) for constant supply voltage and variable impedance.

For a sinusoidal supply using inductors $X_L = 2\pi f L$ Ω or capacitors $X_C = (1/2\pi f C)$ Ω. Reactive components store energy $\frac{1}{2}Li^2$ (inductor), or $\frac{1}{2}Cv^2$ (capacitor), which assists in maintaining the discharge by opposing the change and maintaining the current (inductors) or voltage (capacitors).

The voltage drop across reactive components is 90° out-of-phase with the voltage across a series resistor, leading in the case of an inductor and lagging for a capacitor (Figure 9.4).

The current and voltage can be calculated using complex numbers or by vectors. The impedance Z is the vector sum of the resistance and the reactance of the circuit.

To maintain a stable discharge under a range of conditions such as changes in pathlength, the series impedance of an inductor with a ferrous core may be increased to vary the current and obtain a drooping characteristic (Figure 9.5) by using magnetic shunts on the transformer, or by moving the core apart to divert the flux path and increase the flux leakage or by saturation of the core.

It is not possible to operate more than one plasma from the same supply with a single stabilizing resistor, since their negative dynamic resistance results in all the current passing through one discharge. Any number of discharges can be operated from the same supply provided that they are separately stabilized.

Separately stabilized discharges can also be operated in close proximity so that they coalesce due to the interaction of their magnetic fields forming a single

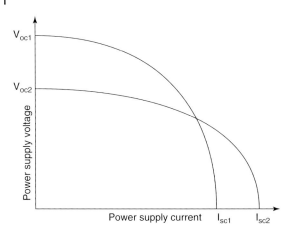

Figure 9.5 Drooping load characteristic.

discharge with separate anode and cathode roots, if both electrodes of each discharge are separately stabilized.

The repulsive force between adjacent electric discharges where the currents have opposite directions (anti-parallel) enables two or more discharges to exist separately close together without coalescence.

In an AC circuit containing resistance, the current and voltage are in-phase. In a circuit containing inductance, the current lags the voltage whereas in a circuit containing capacitance the current in the capacitor leads the voltage. This is shown in Figure 9.6.

A purely reactive component dissipates no power. The total current is the vector sum of the active power component of current and the reactive component which are in quadrature, the power $W = VI\cos\phi$, the volt amps (VA) and the reactive component of power VAR. The ratio W/VA is the power factor $\cos\phi$. where ϕ is the difference in phase between the output voltage and current.

A plasma supplied with AC results in the production of harmonics, which may cause resonance. A simple vector addition of the voltages and current in the circuit may be sufficient, but if the waveforms are non-sinusoidal or contain harmonics, a more detailed transient circuit analysis may be required.

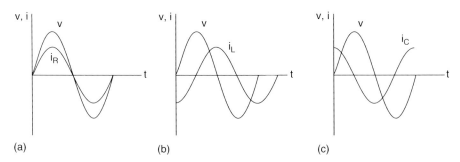

Figure 9.6 Phase shift: (a) resistive, voltage and current in-phase; (b) inductive current lagging; (c) capacitive current leading.

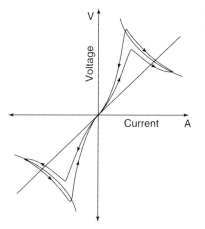

Figure 9.7 Effect of frequency on the plasma dynamic impedance.

9.4
Effect of Frequency

If the supply is AC, the current follows the voltage with a delay of about 1 ms caused by the inertia of the ions along the characteristic in a hysteresis curve (Figure 9.7). The time for decay of an electron in the gap is very short, $\sim 10^{-10}$ s, but the inertia of atoms and ions is much longer, $\sim 1 \times 10^{-3}$ s, and when the duration of half a cycle from the power supply is less than $\sim 1 \times 10^{-3}$ s the persistence of the condition for ionization in the gap assists reignition of the discharge on the next half cycle; the current in the discharge ceases to flow and the dynamic resistance of the discharge becomes positive, corresponding to the resistance of the plasma (without the discharge current) as a shallow curve (as the resistance increases) passing through the origin. The dynamic characteristic becomes a straight line passing through the origin of slope v_D/i_D. All these effects contribute to a sudden change in parameters over the range 50 Hz–1 kHz.

9.5
Interaction between the Plasma and Power Supply Time Constants

The transient behaviour of the electrical supply circuit including the power supply is important in determining the behaviour of the plasma.

If the voltage across a plasma suddenly changes, the effect of the inductor is to try to maintain the current constant by developing a back e.m.f. to oppose the current, $e = -L(di/dt)$; the change in voltage is exponential with a time constant $\tau = L/R$ at which the voltage increases or decreases to $1/e$ of its initial value (Figure 9.8). If the duration of the change is short compared with the circuit time constant, then the current will not vary significantly. The effect of a capacitor is to maintain the voltage across it constant by developing an opposing current $i = -Cdv/dt$ the time constant is CR.

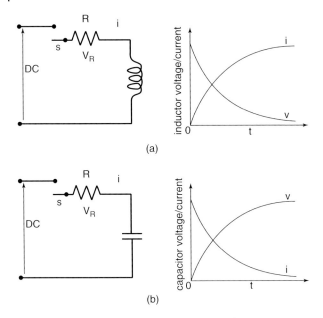

Figure 9.8 Transient behaviour of inductive and capacitive capacitance circuits (a) inductive, (b) capacitive.

The condition for maintenance of a discharge is achieved if the duration of the breakdown pulse τ_{ig} is long enough for the current from the power supply to reach a value sufficient to maintain the discharge, that is $\tau_{ig} > \tau_s$.

The time constant of the power supply circuit includes the resistance of the plasma and will vary during the establishment of a plasma as the plasma resistance decreases. Ignition of the plasma is also affected by the power supply circuit but the circuits are usually decoupled by capacitors and inductors. If the time constant of the ignition supply is short, a series of ignition pulses may be required to build up sufficient ionized species and the supply current builds up sufficiently to maintain the discharge.

If the time constant of the power supply circuit τ_{ps} is longer than the time constant of the discharge τ_d, the power supply will react slowly to any sudden changes caused by fluctuations in the discharge and will tend to maintain the current at a constant value. When τ_{ps} is less than τ_d, the time constant of the discharge, the discharge will tend to follow fluctuations in the discharge current.

9.6
Matching

Matching the supply to the plasma load is a critical part of the power supply operation and is highly dependent on the application. Matching is required to control the power supply output so as to obtain the correct operating conditions of the plasma and the designed power output from the supply. A mismatched power supply may deliver only a fraction of the available power and result in potentially damaging reflected energy and is also necessary to make accurate measurements and avoid reflections due to changes of impedance.

At power frequencies and medium frequencies up to about 100 kHz, transformers can be used where the matching relation between primary and secondary is derived from the transformer relations $V_1/V_2 = N_1/N_2 = I_2/I_1$, which gives

$$Z_p = Z_s \left(\frac{N_p}{N_s}\right)^2 \tag{9.5}$$

For many applications and in particular at power frequency and high currents, the impedance of the power source without an additional matching component is low. At low frequencies, <100 kHz, inductors are widely used for stabilization and matching.

Variable matching is necessary because of the variation in the plasma, such as length, volume, gas, gas pressure and other process parameters. Capacitors can be used to match the load to the supply (Figure 9.9) in different configurations at RF.

Matchboxes use motors to drive air-gapped capacitors to match plasma loads at high frequencies.

Most commercial RF supplies have an output impedance of 50 Ω, which is used as an industry standard for RF coaxial cables. The problem becomes one solely of matching the output voltage and impedance of the supply ($Z_p = 50\,\Omega$) to the load, again using the relation $Z_p = Z_s (N_p/N_s)^2$, where $Z_s = V_s/I_s$, to specify the matching transformer. The effectiveness of matching is given by the voltage reflection factor, $\rho = \frac{Z_L - Z_0}{Z_0 + Z_L}$, which is zero when $Z_0 = Z_L$.

The impedance of a line is matched if no reflection from the end of the line occurs. Reflections occur when a line is not correctly matched. If the reflected wave is out-of-phase with the transmitted wave, the voltages of the transmitted and reflected waves add or subtract and can damage the power supply or cause instrumentation errors if it is not correctly matched to the source or instrument. Over distances $\ll \lambda$, a reflected wave is close to being in-phase with the transmitted wave.

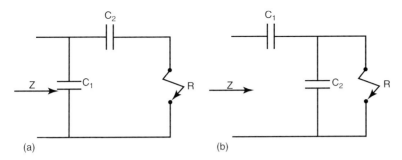

Figure 9.9 Some configurations using capacitors to match RF supplies to plasmas (a) low impedance (b) high impedance.

Microwave waveguides use mechanical tuning stubs to match the waveguide by changing the geometry of the waveguide. Horn-shaped waveguides are used to propagate the microwave into a reactor and match the impedance to free space and avoid reflection. Access to the waveguide can be made via quarter-wavelength choke.

9.7
Resonance

In addition to inductance and capacitance intentionally connected in discharge circuits, self-capacitance between turns in windings of inductors and self-inductance of capacitors, interconnections, electrodes and adjacent conductors, magnetic flux linkage (mutual flux) between current carrying conductors and the leakage reactance of transformers exist.

Examples of series and parallel resonant circuits and a transmission line or connecting cable are shown in Figure 9.10. The inductors and capacitors are shown as discrete components but may be distributed over the length of conductors in

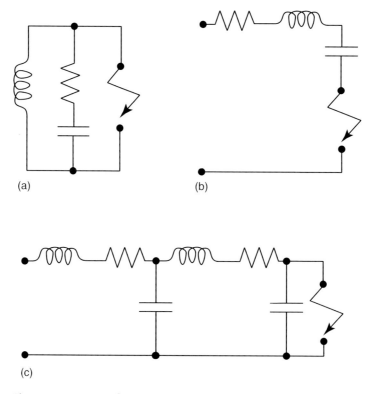

Figure 9.10 Sources of resonance in plasma supply circuits (9/9): (a) series resonance; (b) parallel resonance; (c) leads.

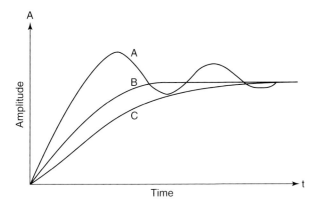

Figure 9.11 Resonance conditions: (a) underdamped; (b) overdamped; (c) critically damped.

a circuit and include parasitic capacitance or inductance and behave more like transmission lines at high frequencies.

$$v = iR + L\frac{di}{dt} + \frac{1}{C}\int i\,dt \tag{9.6}$$

The equation for parallel resonance is similar. The series and parallel equations for resonance have the solutions underdamped (oscillatory), $R < 2\sqrt{L/C}$, critically damped, $R = 2\sqrt{L/C}$, and overdamped, $R \geq 2\sqrt{L/C}$, shown in Figure 9.11.

The resonant frequency for a series circuit is given by

$$f_0 = \frac{1}{2\pi\sqrt{LC}} \tag{9.7}$$

The peak output voltage at resonance becomes $Q\hat{v}$, where $\hat{v}$ is the peak oscillatory voltage. The Q or *magnification factor* of the supply voltage is given by

$$Q = \frac{\omega_0 L}{R} = \frac{1}{\omega_0 CR} \tag{9.8}$$

9.8
Parasitic Inductance and Capacitance

The inductance of a conductor radius separated by a distance d between centres ($d \gg a$) is due to the external flux linkage $\left[\mu \ln(d/a)/\pi\right]$/m^{-1} (Figure 9.12). This is important even at low frequencies, for example in switch-mode power supplies where the peak DC current is very high (CR).

For an air-cored coil (ignoring end effects and for a ratio of the coil length l to the diameter $\gg 1$) the inductance is approximately

$$\frac{N^2 A \mu_0 \mu_r}{l} \tag{9.9}$$

where A is the diameter of the coil and μ_r is the relative permeability.

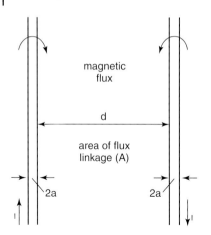

Figure 9.12 Inductance due to flux linkage between parallel conductors.

The capacitance of the discharge circuit may also be significant, particularly at high frequencies. The capacitance for two parallel wires (Figure 9.13) with separation between centres $d = 2h$ in a medium with relative permittivity ε_r is

$$C\,(\mathrm{F\,m^{-1}}) = \frac{\pi\varepsilon_0\varepsilon_r}{\ln\left(\frac{d-a}{a}\right)} \tag{9.10}$$

and for $a \ll d$, $C = \pi\varepsilon_0\varepsilon_r / \ln(d/a)$ F m^{-1} and the capacitance to the earthed plane is

$$C\,(\mathrm{F\,m^{-1}}) = \frac{\pi\varepsilon_0\varepsilon_r}{\ln\left(\frac{d}{a}\right)} \tag{9.11}$$

The capacitance of the electrode gap can also be estimated in a similar way since for two parallel plates of area A and separation d

$$C\,(\mathrm{F\,m^{-2}}) = \frac{A\varepsilon_0\varepsilon_r}{d} \tag{9.12}$$

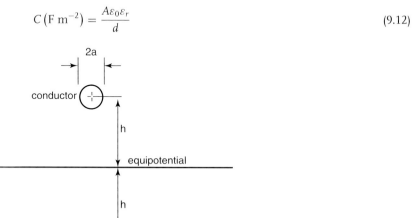

Figure 9.13 Mirror image of a conductor.

Even though the gap may have a capacitance of picofarads, the effect may be apparent at high frequencies.

Where the capacitance or inductance is distributed, for example a wire, it behaves like a transmission line and a single resonance condition does not exist, but where discontinuities occur, for example at connections to discrete components, more than one resonance condition will exist. In some cases the discharge may form part of a resonant circuit, for example where the discharge has an inherent capacitance such as in a barrier discharge and inductance is used to stabilize it.

The transient behaviour of an electric discharge or plasma may be affected by much smaller values of reactance than required for stabilization, including stray (parasitic) inductance due to, for example, the self-inductance and capacitance of leads and even the electrodes. These effects may be very small for plasmas of low impedance, but at low currents where the plasma impedance is high the effects may be sufficient to prevent ignition or may be comparable to the plasma current.

Although inductance or capacitance may not deliberately be connected in the circuit, the effect of stray inductance and the apacitance can have significant effects. The effect of inductance results in sparks on interruption of DC circuits of a few volts and the capacitance of for example welding electrodes, although small, may be sufficient to cause high levels of ultraviolet light when rapidly discharged when the arc is extinguished, which may cause eye damage.

Further Reading

Hughes, E., Hiley, J., Brown, K. and McKenzie-Smith, I. (2008) *Hughes Electrical and Electronic Technology*, 10th edn, Pearson Education, Harlow.

10
Plasma Power Supplies

10.1
Introduction

The power delivered by the supply to a plasma is determined by the dynamic state of the plasma and is affected by factors such as voltage, current, gas pressure, impurities and working gases, making it impossible to analyse the behaviour of the power supply and the plasma independently.

The design requirements of a supply start with the requirements of the plasma, principally power and power density at the operating pressure, gas, and so on. The key plasma parameter is the voltage drop across the plasma and power required, and hence current and impedance to give the required power and power density, working back to matching the supply to the plasma.

The different functions of a power supply are shown schematically in Figure 10.1. The principal power supply components are shown in Figure 10.2.

10.2
Transformers and Inductors

Inductors and transformers are important stabilizing and matching components used in plasma power supplies and are the way in which the voltage, current and power are controlled.

DC plasmas using rectified AC can be stabilized by connecting an inductor in series with the AC input to the rectifier. An inductor may also be connected in the DC supply to smooth the output and to stabilize further the discharge current as opposed to defining the operating point.

Wound components such as inductors and transformers may be air cored or have ferromagnetic (iron or ferrite) cores. A ferromagnetic material provides a lower reluctance path for magnetic flux than air. Steel laminations can be used up to about 10 kHz, when induced eddy current losses due to current circulating in the core become significant.

The two defining equations are Lenz's law, $e = -L di/dt$, and Faraday's law, $e = -N d\phi/dt$ which can be used to express the inductance and voltage since

Introduction to Plasma Technology: Science, Engineering and Applications. John Harry
Copyright © 2010 WILEY-VCH Verlag GmbH & Co. KGaA, Weinheim
ISBN: 978-3-527-32763-8

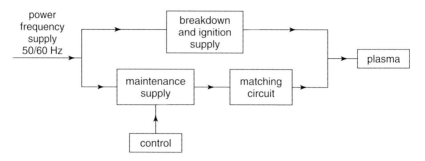

Figure 10.1 The functional parts of a plasma power supply.

$$L = N\frac{d\phi}{di} = N\frac{dBA}{di} = \frac{N^2 A \mu_0 \mu_r}{l} \tag{10.1}$$

$$\hat{V} = -N\frac{d\phi}{dt} = \omega \hat{B} A N^2 \tag{10.2}$$

The inductance depends on the value of the relative permittivity μ_r, which varies with current and the material of the magnetic circuit (Figure 10.3).

A ferromagnetic core has a relative permeability $\mu_r \gg 1$ and confines the flux to the magnetic path $(l/A\mu_0\mu_r)$. Saturation of the core occurs as the current increases and the relative permeability $\mu_r = dB/dH$ decreases (Figure 10.3b), and the inductance decreases to the air-cored value unless otherwise limited.

The effect of even a small air gap in the flux path is to increase greatly the reluctance of the flux path, and reduce saturation. The reluctance of the magnetic circuit becomes $\sum (l_{fe}/A_{fe}\mu_0\mu_r + l_a/A_a\mu_0)$; however, although the total number of ampere turns needed is increased to achieve the same value of inductance,

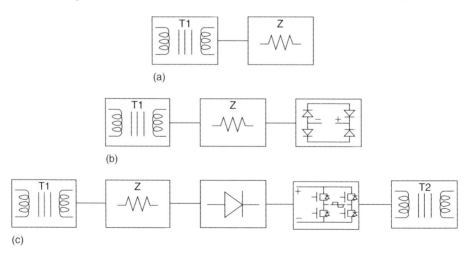

Figure 10.2 Principal power supply types and components: (a) power frequency supply; (b) rectified supply, (c) inverter.

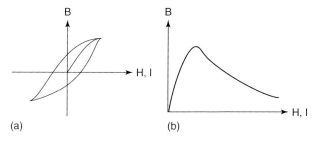

Figure 10.3 Variation of magnetic flux density (B) with magnetic field potential (H) and relative permeability (μ_r) for a ferromagnetic material. (a) B–H loop and (b) permeability.

saturation is reduced. Similarly, an air gap in a transformer core decreases the magnetizing inductance but has little effect on the leakage flux.

Inductors are used to stabilize AC discharges and also to smooth rectified AC and to stabilize welding arcs in DC. When inductors are used in DC circuits, the core has an air gap in the magnetic circuit which prevents saturation. Air gaps are also used to reduce winding inductance so as to allow resonance at higher frequencies.

In addition to air gaps, the inductance can be controlled by the geometry of the windings and the core. Variable inductance can be achieved by mechanically changing the coupling between the core and the winding or by deliberately saturating the core of an inductor with a DC current in the magnetic amplifier or saturable reactor.

Transformers are used in plasma power supplies for isolation from the power frequency supply and also for matching the plasma load to the supply and for producing high voltages for ignition. The ideal transformer is shown in Figure 10.4a and is a good approximation for many applications at power frequency, where the leakage reactances between the windings of the transformers are low, but is not so accurate at high frequencies. The voltage equation for an ideal transformer is

$$\frac{V_p}{V_s} = \frac{N_p}{N_s} \text{ and } I_p N_p = I_s N_s \tag{10.3}$$

where $V_{p,s}$ are the winding voltages and $I_{p,s}$ are the currents.

In the lossy transformer (Figure 10.4b), the windings have resistance (R_p, R_s) and leakage inductance (L_p, L_s). L_m and R_e correspond to magnetization current and eddy current losses.

The leakage flux between windings of a transformer is minimized by winding the primary on top of the secondary, but can be increased by winding the secondary on top of the primary alongside the primary, or on a different limb of the transformer core. High-voltage transformers, in particular, are often wound with the secondary winding on top of the primary to improve the insulation between windings. The flux linking the windings can be changed by using additional core material to shunt the windings so as to divert the flux path in the core to vary the current,

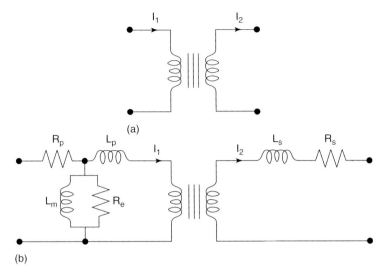

Figure 10.4 Equivalent circuits of the transformer: (a) the ideal transformer; (b) transformer with losses.

or by moving the core apart. A nonlinear drooping voltage–current characteristic (Figure 9.6) can be obtained by saturation of the core.

At high frequencies, the leakage reactance of a ferrite-cored transformer is high. At frequencies above about 1 MHz, air-cored Tesla transformers are sometimes used (Figure 10.5). The air-cored Tesla transformer has a low leakage inductance and can be made to resonate at high frequencies using the intrinsic inductance and self-capacitance of the transformer windings to resonate.

The output voltage is obtained at the resonant frequency $f_0 = 1/2\pi\sqrt{L_t C_c}$, where L_t and L_c are the effective total inductance and capacitance of the transformer, respectively.

The output voltage of a transformer can be varied by changing the number of turns on the secondary with fixed tappings. The auto transformer has a single winding with fixed taps or continuously variable tapping points (Figure 10.6). As it has a single winding it is lighter and cheaper, but it does not provide isolation between the input and output. (When an auto transformer is used at 50 Hz,

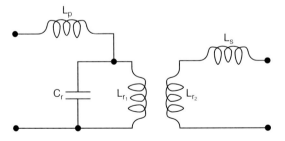

Figure 10.5 Equivalent circuit of the Tesla air-cored transformer.

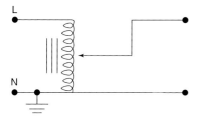

Figure 10.6 The auto-transformer.

the neutral connection which is normally connected to earth at the substation transformer cannot be connected to earth at the transformer without potentially dangerous earth currents flowing. The same problem occurs with switch mode power supplies where no isolating transformer is used.)

The transformer voltage equation $\hat{V} = \omega \hat{B} A N$ is proportional to the supply frequency if everything else is held constant. The use of frequencies higher than the power frequency enables fewer turns and a smaller core to be used. A typical operating frequency is 20 kHz above the upper limit of the audio range (<18 kHz). Transformers can be used to step up or step down the supply voltage but are often limited by practical considerations such as conductor cross-section to ratios of about 40 : 1, and it is often more practical to use semiconductors to step up (boost) or step down (buck) the voltage.

The transformer leakage inductance and winding capacitance can be made to resonate. Resonance at high frequencies (>100 kHz) is possible using ferrite-cored transformers with a small series gap in the core, which reduces the effective inductive reactance. Series resonant circuits allow the transformer output voltage on no load to be increased and permit ignition of a discharge. The circuit magnification factor $Q = \omega L/R$ is high on no load and after ignition drops due to the discharge resistance. The discharge voltage and power can also be varied by varying Q.

10.3 Rectification

Figure 10.7 shows some rectifier configurations and their outputs. The output of a rectifier is a series of rectified pulses; the smoothest output occurs with the highest number of pulses and multiphase rectification is often used so as to reduce the size of smoothing components. The output from the rectifier is connected to large power capacitors to smooth the output. A three-phase supply is often used even at low currents to reduce the inrush current on each conduction cycle so as to reduce the capacitance needed to smooth the output. The value of the capacitance required needs to be sufficient to maintain the current when the rectifier is not conducting and six or even 12 phase rectifiers may be used at high current. The size of the capacitors required is determined from the stored energy $\frac{1}{2}CV^2$ and the acceptable level of decrease in voltage at the end of the conduction cycle. Electrolytic capacitors are normally used, which must have adequate AC current and power ratings.

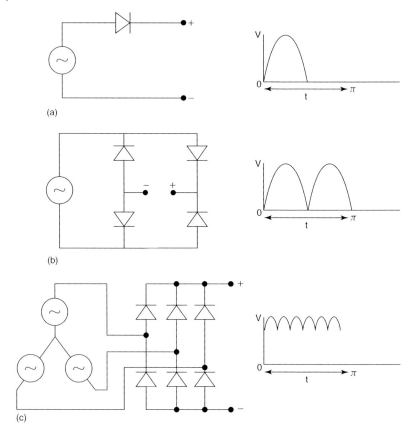

Figure 10.7 Rectifier configurations and their output waveforms: (a) single-phase (single-pulse) full-wave bridge rectifier; (b) full-wave bridge (two-pulse) rectifier; (c) full-wave bridge (six-pulse) rectifier.

With very high peak inrush currents of hundreds or even thousands of amperes, the output is connected to the capacitors by closely spaced or twisted pair conductors; or at very high currents, parallel busbars to minimize the voltage drop by minimizing their mutual inductance so that the inductance is only slightly more than the self-inductance of the busbars. The busbars are designed to minimize the voltage drop due to resistance by using a thickness approximately double the skin depth $\left(\delta = \sqrt{\rho/\pi f \mu_0 \mu_r}\right)$ with sufficient width to reduce the resistance.

Rectifier circuits can also be used as voltage multipliers. Figure 10.8 shows the use of half-wave rectifiers as voltage multipliers to supply a high voltage at low current. Each diode charges the capacitor associated with it, which is connected in a series arrangement so that the output voltage is the sum of the individual voltages. Other similar rectifier configurations also exist.

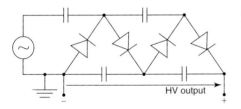

Figure 10.8 Half-wave voltage multiplier.

10.4 Semiconductor Power Supplies

Inverter supplies are simple in concept and used in low-cost consumer equipment such as DC–AC converters, discharge lamps and AC and DC welding supplies. The low thermal inertia of power semiconductors requires more protection against potential fault conditions than most passive components. (Common fault conditions include overvoltage, undervoltage, inrush current, transients, steady state, overcurrent, short circuit and no load.) However, this is often outweighed by increased flexibility, lower material costs and compact design made possible by operation at a higher frequency than the power frequency.

The basic components of an inverter or switch mode power supply (SWMPS) are shown in Figure 10.9. The power input is normally at the power frequency supply voltage without a transformer using either a single- or three-phase rectifier without a transformer and the output is not isolated from the supply input, which may present isolation problems.

10.4.1 The Inverter Circuit

Schematics of full- and half-wave bridge inverters are shown in Figure 10.10. The switches in the full-wave bridge are (a) switched alternately to invert the DC input to give a square-wave output by switching in sequence 1 and 4, 2 and 3 (b). Control can be to two (half-wave bridge) or to all four (full-wave bridge). The effect of the capacitance in the half-wave bridge in series with the supply is to limit the lowest output frequency. For a given DC supply, the output voltage for the full-wave bridge is twice that for the half-wave bridge. The voltage rating of the switching devices is the same for both, but the half-wave bridge cannot operate at low frequencies unless the capacitors are large.

The circuits of full- and half-wave bridge inverters using semiconductor switches are shown in Figure 10.11. Choice depends on the semiconductor rating, frequency and circuit complexity.

Figure 10.9 Principal components in an inverter power supply.

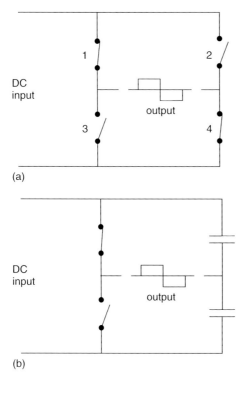

Figure 10.10 Schematics of (a) full and (b) half-wave bridge inverter circuits.

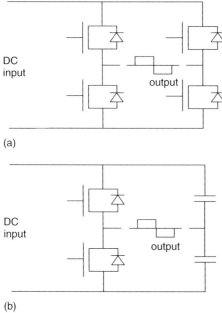

Figure 10.11 Inverter bridges using semiconductors.

Diodes are connected across the semiconductors to provide a path for the reverse current for the back e.m.f. from the switching transients. The function of the diodes across the semiconductor switches is to enable the current to continue to flow as very high voltages would be induced by the inductors if the current were to be instantly cut off. The output is a square wave which has a fundamental frequency the same as the switching frequency and harmonics; a sinusoidal output can be obtained by connecting a filter circuit in series.

10.4.2
Semiconductor Switches

Table 10.1 lists the maximum currents and frequencies of semiconductors used in inverters.

Very high currents can be controlled with thyristors; however, the voltage across a thyristor has to be reduced to zero to turn it off (commutation), which takes about 20 μs and limits their operating frequencies to about 20 kHz. Insulated gate bipolar transistors (IGBTs) and metal oxide semiconductor field-effect transistors (MOSFETs) do not require the voltage across them to be reversed to turn them off and allow faster switching times to be obtained. IGBTs can operate at up to about 80 kHz and carry currents of 200 A and voltages up to 3 kV. The highest frequency of operation is obtained using MOSFETs, which can operate up to 1 MHz, 1 kV and 150 A.

10.4.3
Current Control

The current outputs corresponding to pulse amplitude and frequency control are illustrated in Figure 10.12. An AC output or unidirectional DC output is possible.

The output voltage of the inverter is proportional to the ratio of the on time divided by the cycle time (t_0). A number of different strategies can be used to control the output current and power, including (i) pulse width modulation in which the cycle time is constant and the output is a series of pulses with a variable ratio of on time to cycle time Figure 10.12a or (ii) pulse rate modulation in which the pulse length is constant but the frequency varies.

Table 10.1 Maximum currents and frequencies of semiconductors used in inverters.

Device	Voltage (kV)	Current (A)	Frequency
Thyristor	5	3000	500 Hz
IGBT	3	500	80 kHz
MOSFET	1	150	1 MHz

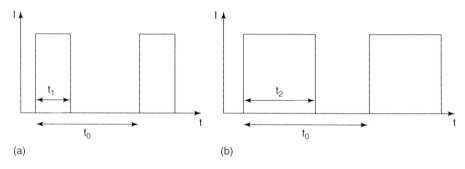

Figure 10.12 Amplitude and frequency modulation of inverter output: (a) amplitude modulation; (b) frequency modulation.

10.4.4
The Inverter Circuit

Figure 10.13 shows a half-bridge control circuit used as a basis of many different plasma supplies. The two MOSFETs or IGBTs are switched alternately at a pulse repetition frequency set by a suitable signal at the input. It is essential that the two MOSFETs do not turn on together even for a very short time. The capacitor C_b in the control circuit is to enable the upper MOSFET to be driven correctly. The two series capacitors effectively establish the output at half the DC supply voltage.

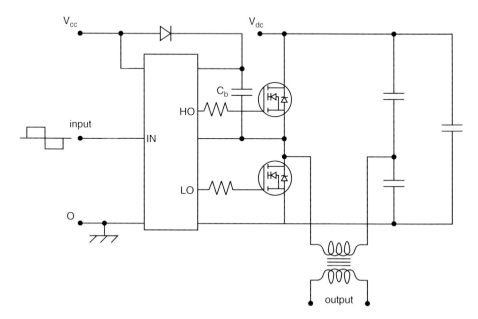

Figure 10.13 Controlled half-bridge inverter using MOSFETs.

10.4.5
Converter Circuits

Converter circuits have the advantage of simplicity and small number of components and the ability to change DC voltages without a transformer. Boost and buck and flyback circuits are used to step voltages up and down as an alternative to transformers when the transformer ratio is high.

Examples of boost and buck supplies and a flyback converter circuits are shown in Figure 10.14. When the switch S is closed, current flows in the inductor (a) and energy is stored. If after a time t S is opened, the inductor discharges through the rectifier and the stored energy in the capacitor builds up. After a few cycles, an equilibrium condition is reached when the energy stored in the capacitor is equal to the energy stored in the inductor. The converters store energy in an inductor to increase (boost) or decrease (buck) the supply voltage. For the boost converter if the output is continuous, $V_{out} = V\,(T/T - t)$, where T is the period. A similar analysis can be made for the buck converter.

The principle of the flyback converter is shown in Figure 10.14c. The flyback converter uses a transformer as an inductor to store energy from the magnetizing current which is released in a tightly coupled secondary winding when the input is open-circuited. The primary is prevented from saturating by an air gap in the transformer core. The output voltage is a function of the turns ratio, and the ratio of the on time to the period t/T:

$$V_{out} = V\left(\frac{T}{T-t}\right)\left(\frac{N_2}{N_1}\right) \quad (10.4)$$

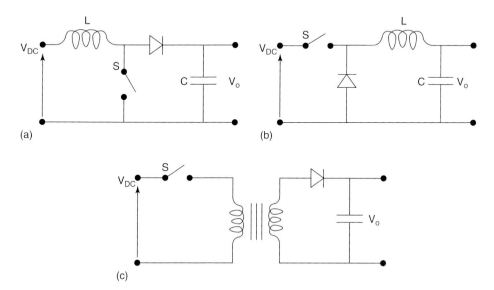

Figure 10.14 Schematic circuit of (a) boost, (b) buck supplies, and a flyback converter (c).

Table 10.2 Some commercially available RF power supply frequencies and powers for electric discharges and plasmas.

Frequency	Power output
10–200 kHz	500 W – 30 kW
450 kHz	1–250 kW
13.56 MHz	200 W – 30 kW
27.12 MHz	300 W – 30 kW
915 MHz	1–50 kW
2.45 GHz	300 W – 6 kW

10.4.6
Inverter Frequencies

Inverters are capable of operating from DC to megahertz frequencies depending on the semiconductor switch and the circuit topology. The output frequencies of commercial plasma power supplies are normally confined to the Industrial, Scientific and Medical (ISM) bands designated by the International Telecommunications Union (ITU). Some typical frequencies of commercially available power supplies and their power output limited by legislation or by commercial availability are listed in Table 10.2.

10.4.7
High-Frequency Inverter

Figure 10.15 shows an example of a simple switching circuit for use at high frequencies due to the high inherent inductance and capacitance of a high-current bridge circuit. The DC input voltage is supplied to the switching circuit from an SWMPS.

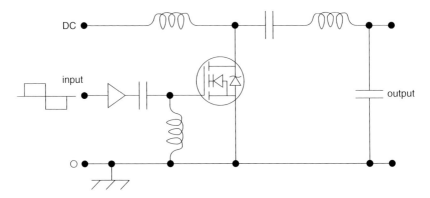

Figure 10.15 High-frequency inverter circuit.

At high frequencies above about 1 MHz and currents of several amperes, the reactance of conductors may limit the resonant frequency and their increase in resistance due to the skin effect may damp the resonant circuit and result in excessive voltage drops.

At frequencies above about 30 MHz, the mutual inductance and capacitance between conductors in a circuit can act as resonant circuits rather than discrete components. A resonant cavity can be formed so that two opposite sides correspond to a capacitor and the other two sides to an inductor.

10.5
Electronic Valve Oscillators

High-power semiconductors for switching are limited to frequencies of a few megahertz and moderate powers (Table 10.1). At higher frequencies electronic valves (electronic vacuum tubes) are used. Figure 10.16 shows a valve oscillator circuit used to drive an inductively coupled high-frequency plasma. The valve uses a beam of electrons from the cathode to the anode, which is switched by the voltage on the grid. Valves can operate at several kilovolts and currents >100 A at frequencies of several megahertz at high powers (>10 kW).

10.6
Microwave Power Supplies

As the frequency required increases to about 1 GHz, the time taken for an electron to travel between the cathode and anode of the valve becomes comparable to the output frequency and magnetrons which are capable of switching at much higher frequencies are used.

A simple microwave power supply is shown in Figure 10.17. The heated cathode acts as a source of electrons which form a beam and are attracted to the anode in an evacuated tube. The output is radiated into a waveguide by an aerial and propagated as a lossless wave in a waveguide (Figure 10.18).

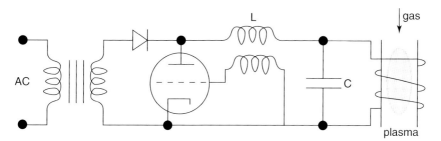

Figure 10.16 High-frequency valve oscillator circuit.

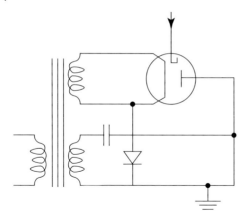

Figure 10.17 Magnetron supply.

The circulator is used to prevent reflected power damaging the magnetron by dumping it in the dummy load. Matching and tuning waveguides may also be used.

10.7
Pulsed Power Supplies

The peak pulse power and energy for a square-wave pulse are related by energy $\xi = W\tau$. The duration of the pulse should be sufficiently short to enable a high electric field to be maintained for the duration of the discharge so as to prevent charge equilibrium and allow high-energy ionized species to take part in nonthermal chemical reactions. To prevent thermal equilibrium, the pulse duration is normally less than 1 ms.

To obtain a short pulse length, the rise time and decay should be as short as possible unless a separate series triggered spark gap is used.

The stored energy in a capacitor is $\tfrac{1}{2}CV^2$ and the time constant is RC (the time required for the change to $1/e$ of the original value) so that large amounts of energy

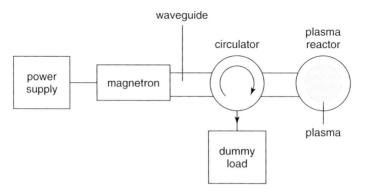

Figure 10.18 Waveguide applicator.

can be stored at high voltages requiring a small resistance in the discharge circuit and large capacitance C. To obtain high currents and rapid current rise times, the inductance of the discharge circuit must be minimized and parasitic inductance minimized.

Resonance of the plasma circuit may occur since any inductance in the circuit will be likely to resonate with the capacitance since, by design, R is small. $R \geq 2\sqrt{L/C}$, the resistance of the plasma may be low at high current and the resistance of the circuit will be low to obtain high peak short duration current pulses. The resonant frequency is $f_0 = 1/2\pi \sqrt{LC}$.

In the case of repetitive pulses, the charging time should be sufficient to minimize inrush but short enough if repetitive pulses are used and the output pulse to be matched to the plasma time constant.

10.8
Ignition Power Supplies

If we consider the initial breakdown process, the breakdown voltage can be supplied by the supply by either a conventional step-up transformer, a resonating transformer (which stops resonating on load), a separate ignition supply or drawing the electrodes apart. The use of a conventional step-up transformer is very wasteful of the transformer kilovolt amperes and seldom used, and drawing a discharge is seldom practical, so we will only consider a transformer with a separate ignition supply where the open-circuit output voltage is twice the plasma terminal voltage.

Ignition of a plasma or electric discharge can be achieved in several ways. The simplest method is to draw a discharge between electrodes by moving them apart. Originally used in lighting applications, it still remains a practical solution to ignite an arc and is used for manual and metal inert gas (MIG) arc welding, arc furnaces and other arc applications. Drawing a discharge has the advantage of simplicity and positive action and the maximum supply voltage need only be sufficient to operate the discharge, which is more than twice the plasma voltage.

In the case of a high-current plasma arc, it is uneconomic to operate with a voltage, that is high enough to break down the electrode gap because of the high kilovolt amperes required. Accordingly, the slope of the load line is much lower, close to the limit where $V_{oc} = 2V_d$ or for AC the vector sum $\overline{V}_s = \overline{V}_d + \overline{I}_d Z$, and the supply has to be ignited or drawn, or for low-current applications the kilovolt amperes value is not so critical and the voltage open circuit may be several times the discharge voltage and the discharge descends to the operating point without passing through the transitions.

Where it is not possible or practical to draw the discharge or plasma, a separate ignition supply connected in series or parallel can be used. This is not required to carry the full load current and enables the plasma supply voltage to be limited to about twice the plasma voltage.

At low plasma currents, the increased cost and size of the supply transformer needed to produce sufficiently high voltage to ignite the plasma and the increased impedance needed to stabilize the plasma are not important and the functions of ignition and supplying the load current can be combined.

The load lines for power supplies with the same open-circuit voltage but different series resistance are shown in Figure 10.19. By using a separate ignition supply, the power supply voltage can be reduced, decreasing the ratio of open-circuit voltage to plasma voltage and hence utilization (cost) and safety.

Single-pulse ignition supplies are frequently used for applications such as high-pressure lamps and lasers. In its simplest form shown in the circuit in Figure 10.20, a capacitor is charged to the breakdown of the discharge voltage.

A capacitor charged to a high voltage from a DC supply forms the basis of simple ignition supplies for intermittent use, for example for discharge lamps. However, the initial peak current from the capacitor may cause damage to the electrodes if frequently used. A transformer or half-wave voltage doubling circuit is usually required to step up the power supply voltage before it is rectified. The output is potentially capable of giving a fatal electric shock from the high DC current and stored energy and discharge resistors must be connected. However, the time constant RC of the charged capacitor and discharge circuit is usually

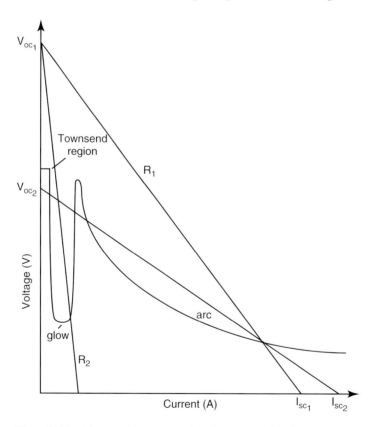

Figure 10.19 Schematic showing supply voltage with and without separate ignition supply.

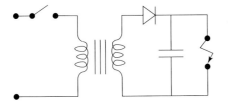

Figure 10.20 Simple capacitor ignition circuit.

several seconds and the output may be dangerous even if discharge resistors are fitted to the capacitor.

Reignition of high-pressure discharges, such as lamps when hot, requires higher voltages than from cold, which may be acceptable for applications where they are allowed to cool before reignition, but is unacceptable for other applications which are switched frequently, for example for car headlamps.

High voltages can be generated by resonating transformers in a similar way to spark plug ignition circuits by switching the current in the primary of a resonating (underdamped) transformer. A resonant ignition circuit is shown in Figure 10.21. The thyristor switches the current in the primary circuit of an iron-cored auto-transformer with a high-voltage secondary, similar to a car ignition coil. The output frequency is determined by the transformer leakage inductance and the capacitance and is typically several hundred kilohertz.

Where safety considerations dictate, such as in tungsten inert gas (TIG) welding, high-frequency ignition is used because the high frequency has less of a tendency to interfere with the rhythm of the heartbeat. The output is normally a high-voltage, high-frequency train superimposed on the power supply voltage either in series or parallel with the electrodes or as an external trigger source. The elimination of the high-voltage DC from the charged capacitor and use of high frequency, which tends to produce burns rather than stopping the heartbeat, are a safer alternative.

Very high voltages can easily be generated using an air-cored Tesla transformer, which can be designed to have a low inductance and therefore resonate at a high frequency and the transformer does not saturate (Figure 10.22). The low-inductance transformer windings and the capacitor form a high-Q resonant circuit, which can be used to generate a high-voltage and high-frequency (100 kHz–10 MHz) output to ignite a discharge. The capacitor is charged to a voltage which depends upon the supply voltage and the setting of the spark-gap. When the spark gap breaks down,

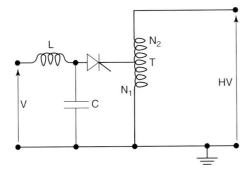

Figure 10.21 Thyristor ignition circuit.

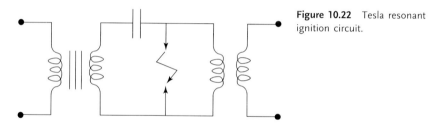

Figure 10.22 Tesla resonant ignition circuit.

C_1 discharges and high-frequency damped oscillations are produced in the primary circuit of the transformer. Although largely replaced by high-frequency electronic ignition, the Tesla transformer is robust and the circuit simple.

In some applications, an ignition voltage higher than the breakdown voltage is required. This has the advantage of a more rapid breakdown and reduces leakage current losses in the circuit, reducing voltage drops.

If a series spark gap is connected in the supply circuit, the voltage at breakdown is increased (Figure 10.23). For a DC voltage, prior to breakdown the voltage is divided between the discharge gap and spark gap.

At high frequencies, the voltages across the spark gaps is inversely related to the capacitances of the spark gap C_1 and electrodes C_2. If the capacitance of the spark gap C_1 is small compared with the capacitance of the discharge electrode gap C_2, prior to breakdown the current will be greater in the spark gap, which will break down first. Alternatively, if the electric field strength of the spark gap is higher than that across the main plasma electrodes, for example by quenching the electron flow by proximity of the walls of the spark gap or by using a gas with a high breakdown voltage, then when the gaps which are in series break down the voltage across the plasma gap is proportionately increased and a higher energy pulse is obtained. When the spark gap breaks down first, almost the entire voltage is available across the plasma electrodes. Operation with an overvolted discharge with a series gap results in more rapid rise times and reduces corona and capacitive losses from leads and leakage currents across the electrodes.

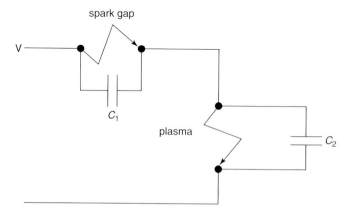

Figure 10.23 Series spark gap ignition.

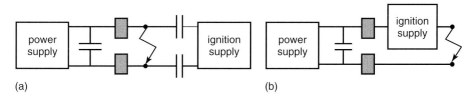

Figure 10.24 Methods of connecting ignition supplies (a) parallel (across) the supply, (b) series connection.

The ignition supply may be connected in parallel or series with the supply (Figure 10.24) depending on the ouput impedance of the ignition supply. With direct connection to the electrodes or for electrodeless plasmas, a third or trigger electrode may be used, or a separate external wire relying on the capacitive coupling to earth. Decoupling the ignition supply from the power supply is relatively easy since the rise time of the ignition pulse is usually very rapid with ferrite rings on the cables to the power supply and capacitors on the leads from the ignition supply.

The electrode circuit may affect the output pulse voltage and waveform due to its inductance and, in particular, the capacitance between the leads to the electrodes both from the ignition supply and from the power supply.

10.9
Electromagnetic Interference

Plasma and electric discharges produce electromagnetic interference (EMI) due to their negative dynamic resistance which causes resonance in underdamped circuits (Figure 10.25). For critical damping (see Section 9.7), the resistance of the plasma $R_d = 2\sqrt{L/C}$, which is not normally acceptable because of the excessive power loss.

EMI can cause malfunction of instrumentation and control equipment and in extreme cases potentially fatal accidents. Legislation exists to limit EMI (IEC Standard IEC 50(161), BS 4727). The standard defines EMI as 'the ability of a device, unit of equipment or system to function satisfactorily in its electromagnetic environment without introducing intolerable electromagnetic disturbances to anything in that

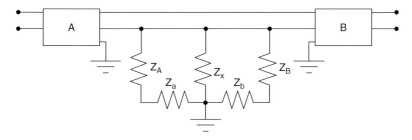

Figure 10.25 Transmission paths of EMI.

environment'. EMI can be limited either by containing the interference in an environment in which it is acceptable or by limiting it at source.

EMI can be coupled by conduction, induction (magnetic coupling) due to a change in current and capacitively (electric field coupling) due to a change in the electric field, and by radiation or air-borne coupling by propagation of an electromagnetic field. Electric coupling tends to be more of a problem in high-impedance circuits and magnetic coupling in high-current circuits. Inductively and capacitively coupled EMI usually occur in the near field; however, airborne coupling occurs in the far field and may travel a considerable distance.

EMI is often caused by resonance, which can be reduced by damping using additional resistance, but this often results in an unacceptable power loss; if it is possible, the resonant frequency can be changed by the choice of circuit inductance and capacitance or by minimizing stray values of inductance or capacitance to reduce the effect of EMI from the source or to make it less intrusive.

10.9.1
Conduction

Conduction of EMI occurs when there is a connection, however tenuous, between the source and the victim. The most obvious source of EMI is where the source and the affected equipment are supplied from a common supply such as the power frequency supply. The power frequency conductors can act as transmission lines over long distances and propagate airborne interference, the connecting cables act as aerials.

The use of transformers to isolate the equipment, filters in supply lines, series capacitors or inductors to decouple high-frequency supplies and parallel capacitors to provide a path to earth can be added to reduce conduction of interference to acceptable levels. Screened leads and power supply conditioning can also help.

Two modes of EMI due to conduction exist. In common mode or common impedance coupling, the EMI is in-phase (in the same direction) in two or more conductors. Common mode interference is caused by airborne interference that induces a current in the loop (current in the same direction) including the wires or cable ground plane and impedances connecting the loop to the ground plane. Differential mode coupling (go and return) occurs when the EMI is out-of-phase. In the differential mode, two conductors in the circuit have different mode interference even if they are closely spaced. Common mode EMI can often be attenuated simply by passing the conductors through one or more ferrite rings or beads to increase their inductance, but differential mode EMI is not attenuated and separate rings for each conductor are necessary.

Shared earth connections can lead to interference and can be reduced by coupling at a common point or a low-impedance earth plane can be installed so that the common impedance between the circuits is as low as possible. The impedance is important in addition to the earth resistance since even a small reactance due to the inductance of the earth connection can result in an induced back e.m.f. Ldi/dt.

The voltage induced between two circuits close together by magnetic coupling can be written $e = -M di/dt$, where M is the mutual inductance between the two circuits. The simplest solution is to minimize coupling by positioning the conductors close together and provision of adequate screening. Electric coupling is due to the capacitance between conductors and to earth, $V = C dV/dt$, where C is the capacitance between the circuits and the impedance to earth. If one or both of the circuits are not earthed (floating), there is still a series circuit via the capacitance to earth. Electric coupling decreases as the separation between conductors or earth increases while magnetic coupling increases with the separation.

Airborne radiation is relatively easily dealt with in most cases and may not even be a problem where the source is operated in a specified electromagnetic environment. Where it is not acceptable, appropriate protection includes screening by enclosure in a Faraday cage (in practice an earthed metal enclosure) with all conductors entering or leaving it appropriately decoupled. Sufficient separation between the source of EMI and the victim such that the EMI is sufficiently attenuated is needed.

Further Reading

Conrads, H. and Schmidt, M. (2001) Plasma generation and plasma sources. *Plasma Sources Science and Technology*, **9** (4), 441–454.

Horowitz, P. and Hill, W. (1989) *The Art of Electronics*, 2nd edn, Cambridge University Press, Cambridge.

Raizer, Y.P., Schneider, M.N. and Yatsenko, N.A. (1995) *Radio-frequency Capacitive Discharges*, CRC Press, Boca Raton, FL.

Roussy, G. and Pearce, J.A. (1995) *Foundations and Industrial Applications of Microwave and Radio Frequency Fields*, John Wiley & Sons, Ltd, Chichester.

Schmidt, M. and Conrads, H. (2001) Plasma sources in *Low Temperature Plasma Physics* (eds R. Hippler, S. Pfau, M. Schmidt and K.H. Schoenbach), Wiley-VCH Verlag GmbH, Weinheim, pp. 283–304.

Index

a
abnormal region 47, 58, 80
absorption spectroscopy 161
actinometry 160
ambipolar diffusion 9, 20
anode 48, 57f
– nickel 55
anode fall region 58
APD *see* atmospheric pressure discharge
arc blow 125
arc discharge
– comparison to glow discharge 60
– effect of rotating an 60
arc gas heater 138f
arc interrupter 145f
arc welding 124f
– submerged 129
atmospheric pressure discharge (APD) 103ff
atomic fusion 4
aurora borealis 4

b
ball-lightning 151
barium oxide 65f
barrier discharge 117f
boost converter 197
boronizing 100
boundary collision 25
breakdown voltage-gas pressure relation 41
buck converter 197

c
capacitance circuit, transient behaviour of 180
capacitive coupling 46, 64f, 81
– electrodeless, in a gas laser 92

capacitively coupled plasma (CCP) 29, 64f
– comparison to ECR plasma 86
capacitor ignition circuit 203
carbonitriding 100
carburizing 100
catalyst recycling 134
cathode 48ff
– aluminium 55
– cold 55f, 89, 132
– copper 55
– for arc welding 126
– hollow 82, 138
– hot 55, 89
– root 50, 57
– strontium barium titanate 55, 89
– thermoionic arc 89
– tungsten 127f, 130
– work function 52
cathode fall region 48ff, 56f
CCP *see* capacitively coupled plasma
charge equilibrium 12
charge imbalance 36, 50
charge perturbation 34
charge transfer 24
charged particle
– behaviour in a magnetic field 37
– behaviour in an oscillating electric field 32
– electromagnetic forces on a 31f
– magnetic moment 40
– motion of, in a magnetron 40
Child's law 94
circuit breaker 145f
– gas blast 146
– using sulfur hexaflouride 146
– vacuum 147f
CO_2 laser 93

Introduction to Plasma Technology: Science, Engineering and Applications. John Harry
Copyright © 2010 WILEY-VCH Verlag GmbH & Co. KGaA, Weinheim
ISBN: 978-3-527-32763-8

coil probe 159
cold plasma 1, 3
– low-temperature non-thermal 3
– low-temperature thermal 3
collision between particles 8f
collision cross-section 19ff
collision energy 8f
collision frequency 9, 12, 18f, 30, 32
colour rendering index 141
common mode coupling 206
conduction of EMI 206
contactor 146
converter circuit 197f
corona discharge 47f, 104ff
– application of 110, 113
– burst pulse 106
– glow 106
– on power lines 108f
– pulseless 106
– sheath 109f
– streamer 106
– Trichel pulse 106
Coulomb force 9
coupled charges 29
coupling of electrical energy to a plasma 29ff
coupling process 29, 45ff
current fluctuation 166f
current measurement 170f
current transducer 170f
current transformer
– Rogowski coil 171
cut-off frequency 69, 74

d

DC arc furnace 134f
DC-plasma diode 80
Debye radius 10ff, 35ff
degree of ionization 34
– as a function of temperature 10
dielectric barrier discharge 104, 114ff
– application of 117f
– electrode configuration 115
differential mode coupling 206
diffusion charging 110
dip transfer 126
direct coupling 45ff, 81
discharge column 57f
discharge voltage 49
discharge voltage-current characteristic 174
discontinous discharge *see* partial discharge
dissociation 24
drift velocity 15ff, 32

dust precipitation using electrostatic charging 110
dynamic plasma impedance 179

e

ECR *see* electron cyclotron resonance
elastic collision 9, 15ff
electric arc 123
electric arc melting 131
electric arc smelting 135
electric discharge augmented fuel flame 139
electrical conductivity 17f
electrical discharge 37, 45
– cold nonequilibrium 48
– initiation of 41f
– low-pressure electrical discharge lamp *see* plasma lamp
– natural phenomena 150f
– self-regulating effect 49
– similarity of 42
– voltage-current characteristic 47
electrical measurements 165f
– instrumentation 165ff
– using an oscilloscope 168f
– using probes 168f
electricity generation by nuclear fusion 149f
electrode
– surface current density 49ff
electrodeless reactor 46
electromagnetic field
– behaviour of charged particles in a 31
– distribution of 69
– plasma interaction 63f
electromagnetic interference (EMI) 205ff
electron beam 94f
– high-power 97f
– melting furnace 98f
electron beam evaporation 94
electron cyclotron 4, 72ff
electron cyclotron reactor 85ff
– comparison to RF capacitively coupled plasma 86
– distributed electron cyclotron reactor (DECR) 85
electron cyclotron resonance (ECR) 68, 71f
– application of 74
electron gun 94
electron gyro frequency 72
– relation to magnetic flux density 38
electron mobility 17
electron motion in crossed electric and magnetic fields 39f
electron number density 2, 12, 65, 71

electron plasma resonant frequency   30, 35, 72
electron temperature   2f
electron temperature- gas pressure relation   58
electronic valve oscillator   199
electrostatic charging   110ff
– application of   110
electrostatic deposition   113f
electrostatic precipitator   107, 110f
EMI *see* electromagnetic interference
emission process   51ff
– current densities of   56
energy of atomic transitions   8
energy of molecular transitions   8
excimer   88, 93
excimer lamp   143
excitation process in lasers   92f, 144f
excitiation   22

f
Faraday's law   187
field emission   54
filamentary discharge   54, 114
film deposition   78, 80, 83, 94
fluorescent lamp   59, 88
flyback converter   197
free electron beam   2, 94ff
frequency effect on the plasma dynamic impedance   179

g
gas laser   91ff
gas
– heat content- temperature relation   59
– transport properties of   6
glow discharge   34, 49ff, 52
– colours of gases and vapours in   89
– comparison to arc discharge   60
glow discharge diode   80ff
glow discharge surface treatment   99f
glow region   47f
ground state   21
gyro frequency   38
gyro radius   38

h
Hall effect probe   159, 171
helical reactor   86f
helical resonator   68f, 86
helicon reactor   4, 74f, 87
helicon wave   74

helium neon laser   92
HID lamp *see* high-intensity discharge lamp
high-frequency inverter   198
high-intensity discharge (HID) lamp   141f
– halide   143
high-intensity focused cathode spot   142
high-pressure discharge lamp   141
– in continuous YAG lasers   143
– short-arc DC high-power arc lamp   143
hollow-cathode effect   82
hot plasma   1, 3
– high-temperature   3
hybrid discharge   51

i
ICP *see* inductively coupled plasma
IGBT *see* insulated gate bipolar transistor
ignition circuit   203f
ignition power supply   201f
– connection of power and ignition supply   205
impedance   177
indirect coupling   62ff
induction coupling   62ff
– electrodeless, in a gas laser   92
induction plasma reactor   84
inductively coupled arc discharge   139ff
– application of   141
inductively coupled plasma (ICP)   29, 62f, 84, 199
inductor   177, 183, 187f
inelastic collision   15, 21ff
insulated gate bipolar transistor (IGBT)   195f
interferometry   162f
inverter circuit   193, 196
– current control of   195f
inverter frequency   198
inverter supply   193, 198
ion beam evaporation   94
ion beam plating   96
ion beam process   94
ion etching   80f, 95
ion excitation   145
ion implantation   78, 96
ion laser   144f
– argon   144
– helium-cadmiun   145
ion resonant frequency   9
ion sheath   10
ion temperature   5
ionization gauge   155
ionization potential   26

ionization process 23f
ionosphere 4

k
kinetic plasma 4

l
Landau damping 67, 75
Langmuir probe 156ff
Langmuir-Child equation 53
Larmor radius *see* gyro radius
laser-induced fluorescence (LIF) 160
leakage flux 189
Lenz's law 187
lightning 150ff
– propagation of lightning strikes 151
load line 175
loss mechanism of collision 35
low-pressure discharge
– in semiconductor fabrication 77f
– non-equilibrium cold 77ff
low-pressure glow discharge
– properties of 78
low-pressure glow discharge lamp *see also* plasma lamp 88ff
low-pressure plasma 79

m
magnetic constriction 50
magnetic field 37f
– interaction with a discharge or plasma 59ff
– toroidal 149f
magnetic flux density 38, 71f
magnetic moment of a charged particle 39
magnetic probe 158f
magnetized plasma 37f
magnetoplasmadynamic (MPD) 4
magnetoplasmadynamic power generation 149
magnetron 40, 41, 83f
– motion of charged particles in a 40, 83
– RF 84
mass spectrometry 165
matching of power supply to the plasma load 180f, 189
Maxwell distribution function measurement 156
Maxwell equation 6
Maxwell velocity distribution 6f
Maxwell-Boltzmann energy distribution 7
McLeod gauge 155
mean free path of gases 16, 20

mercury discharge lamp 142
metal inert gas (MIG) welding 124, 126f
metal oxide semiconductor field-effect transistor (MOSFET) 195f
metal surface treatment using glow discharge process 99ff
metastable ionization 25
metastable neutral ionization collision *see* Penning ionization
metastable process 22f
microdischarge 114, 121
microelectronics fabrication 77f
– plasma sources used in 80
microwave interferometer 163
microwave power supply 199f
microwave waveguide 69, 182
microwelding 129
migration velocity of electrostatically charged particles 111f
molecular laser 93
MOSFET *see* metal oxide semiconductor field-effect transistor
MPD *see* magnetroplasmadynamic

n
near field 29ff, 62f, 73, 206
negative dynamic resistance 173, 177
negative ionization 24
neon lamp 2, 91
neutral particle density 155f
nitriding 100
noble gas 16, 22
– energy characteristics of 26
non-equilibrium discharge 104
non-equilibrium plasma
– atmospheric 2, 114
– low-pressure 2
nuclear fusion 149f
number density 155f

o
optical emission spectroscopy 159
optical spectrometer 160
oscillating electrical field 32
oscilloscope 168f
ozone generator 116

p
parasitic capacitance 183f
parasitic inductance 183f
partial discharge in insulation material 118
PCVD *see* plasma chemical vapour deposition

PECVD *see* plasma-enhanced chemical vapour deposition
Peek's equation   109
Penning effect   92
Penning gauge   155
Penning ionization   21ff, 88, 92
Penning reaction   145
photocopying process using corona discharges   113f
Pirani gauge   155
plasma
– application of   2f, 12
– characteristics of   2f, 11f, 173f
– characteristics of thermal arc   123f
– classification of   1
– definition of   1
– effective permittivity   67
– electron frequency of   13
– formation of   15
– initiation of   40f
– load characteristic   178
– observable spectra   159
– parameters of   155
– phase shifts   178
– resistance   173
– single-turn   62
– stabilized   173
– statistical behaviour   6
– toroidal   62ff, 139
plasma behaviour   30ff, 47
plasma boundary   10
plasma chemical vapour deposition (PCVD)   85
plasma diagnostics
– current   170f
– current at high voltage   167
– current density   158
– current fluctuation   165f
– discharge motion   158
– electron density   156, 158, 161
– electron temperature   156, 161
– gas temperature   156
– neutral particle density   155f
– phase shift between plasma voltage and current   170
– pressure gradient in a plasma   163
– scattering measurement   161f
– voltage at high current   167
– voltage fluctuation   165f
plasma display chanel   116
plasma etching   79f
– RF etching   82
plasma fluctuation   175
plasma frequency *see* plasma resonance

plasma ignition   180, 201
plasma lamp
– cold cathode   91
– electrodeless   91
– hot cathode   89
– low-pressure actinic lamp   90
– low-pressure mercury vapour   88ff
– mercury-free   118
plasma melting furnace   136f
plasma model
– simple ballistic   5f
– statistical   5ff
plasma power density   45, 57, 77, 99, 125, 149
plasma power supply   188
plasma process
– elelctron number density of   12
– energy of   3
– in electronics fabriction   77ff
– potential of   3
– temperature of   12
plasma resonance   11, 30, 34f
– relation to the Debye radius   37
plasma resonant frequency   65, 72
plasma screening   10
plasma spraying   130
plasma stabilization   174f
– of AC plasma   176, 189
– of DC plasma   176, 187
plasma surface treatment   99f
plasma torch   4, 59, 100, 129f
– application of   130
plasma-electromagnetic field interaction
– at AC frequencies   29
– at low DC frequencies   29ff
– in the near field   29f
– propagation of electromagnetic waves   65f
– reflection of waves at an interface   66
plasma-energy coupling   45ff
plasma-enhanced chemical vapour deposition (PECVD)   78
plasma-power supply time constants interaction   179f
Poisson's equation   35
power supply frequence   181f
pre-exponential Sommerfield factor   55
probe impedance   169
propagation of electromagnetic waves   65ff, 74, 206
propulsion in space   100f
pulsed discharge   103
pulsed power supply   200f

q

quadrupole mass spectrometer   165f
quasi-continuous fluid   5

r

radiative process   22
radiofrequency plasma (RF plasma)   33, 96
reactive plasma stabilization   176f
recombination process   23f
rectification   191f
rectifier
– full wave bridge (six-pulse)   192
– full wave bridge (two-pulse)   192
– half-wave   192
– single-phase full wave bridge   192
relative permittivity   188
remelting of high-melting metal   137
resistance   176, 178
resonance   182, 191
resonance Landau damping   35
resonance source   182
RF induction plasma torch   139f
RF inductive coupling   46
RF plasma *see* radiofrequency plasma
RF power supply   181, 198
Roussy-Pearce equation   69

s

scattering measurement   161f
Schottky effect   54
secondary emission   52, 82, 94
secondary emission coefficient   53
self-magnetic field   50, 60, 123, 133
semiconductor fabrication   77f, 82
semiconductor power supply   193
semiconductor switch   195
separation of low-density insulating particles   113
series spark gap ignition   204
sheath effect   10, 36f, 51, 105
silent discharge *see* dielectric barrier discharge
similarity parameter   43
skin depth   64
skin effect   65, 68
sodium vapour lamp   142
solenoid   40f, 63, 73f, 85, 87, 145, 171
space plasma   4
sparking potential   41
spectra observed in a plasma   159
spray coating of charged material   113
spray transfer   126
sprite   151
sputtering process   54, 78, 80, 94
stabilization circuit   176
steady-state resistance   174
Stefan-Boltzmann equation   55
strontium oxide   51
sulfur lamp   143
supply   32f
supply frequency   33, 71
supply voltage   174
surface discharge at high voltage   120
surface treatment using barrier discharges   118
switch mode power supply (SWMPS) *see* inverter supply

t

tangential velocity   37
TE mode *see* transverse electromagnetic mode
technological plasma   5
thermal diffusion   6f, 18, 99
thermal equilibrium   115, 123ff, 201
thermal velocity   18
thermionic emission   54f, 94, 132
thoriated tungsten   55
three-phase AC arc furnace   131f
thunderbolt   151
thyristor   195
thyristor ignition circuit   203
Tokamak fusion reactor *see* toroidal fusion reactor
toroidal discharge   91
toroidal fusion reactor   6, 73, 149f
Townsend coefficient   41f
Townsend equation   41
Townsend region   47, 105
transformer   181, 187f
– auto transformer   191
– equivalent circuit of   190
– resonating   201
– step-up   201
transverse electromagnetic mode (TE mode)   69
transverse magnetic field   38
tungsten inert gas (TIG) torch   128
tungsten inert gas (TIG) welding   127f

u

UV light source   90f

v

vacuum arc furnace   137f
vacuum circuit contactor   147f
vacuum deposition   52
Van der Graaf high-voltage generator   151

velocity distribution   6f
voltage fluctuation   166f
voltage multiplier   192

w
waveguide   46, 199, 200
waveguide applicator   200
waveguide coupled plasma   71f

weakly ionized gas   15ff
welding using high-power electron beams   97
work function of metal oxides   55

y
YAG laser   144